T0291827

CAMBRIDGE LIBRARY COLLECTION

Books of enduring scholarly value

Earth Sciences

In the nineteenth century, geology emerged as a distinct academic discipline. It pointed the way towards the theory of evolution, as scientists including Gideon Mantell, Adam Sedgwick, Charles Lyell and Roderick Murchison began to use the evidence of minerals, rock formations and fossils to demonstrate that the earth was older by millions of years than the conventional, Bible-based wisdom had supposed. They argued convincingly that the climate, flora and fauna of the distant past could be deduced from geological evidence. Volcanic activity, the formation of mountains, and the action of glaciers and rivers, tides and ocean currents also became better understood. This series includes landmark publications by pioneers of the modern earth sciences, who advanced the scientific understanding of our planet and the processes by which it is constantly re-shaped.

Catalogue of the Type Fossils in the Woodwardian Museum, Cambridge

The collections of fossils housed in this museum, now known as the Segdwick Museum of Earth Sciences, are of international importance. The original collection was begun in 1728, and grew rapidly. This catalogue by Henry Woods (1868–1952), a graduate of the University of Cambridge who undertook curatorial work in the museum between his graduation in 1890 and his appointment as a Demonstrator in Paleobotany in 1892, was first published in 1891. It contains the specific names, classes and orders of 558 specimens in the museum which are 'type specimens' for particular species, and was primarily intended for scholars searching for the location of those specimens. Woods also included the names of individuals who had described each specimen, the name of the collection it was in and references for the specimen. His book provides a valuable record of important fossils in the collection at the time of publication.

Cambridge University Press has long been a pioneer in the reissuing of out-of-print titles from its own backlist, producing digital reprints of books that are still sought after by scholars and students but could not be reprinted economically using traditional technology. The Cambridge Library Collection extends this activity to a wider range of books which are still of importance to researchers and professionals, either for the source material they contain, or as landmarks in the history of their academic discipline.

Drawing from the world-renowned collections in the Cambridge University Library, and guided by the advice of experts in each subject area, Cambridge University Press is using state-of-the-art scanning machines in its own Printing House to capture the content of each book selected for inclusion. The files are processed to give a consistently clear, crisp image, and the books finished to the high quality standard for which the Press is recognised around the world. The latest print-on-demand technology ensures that the books will remain available indefinitely, and that orders for single or multiple copies can quickly be supplied.

The Cambridge Library Collection will bring back to life books of enduring scholarly value (including out-of-copyright works originally issued by other publishers) across a wide range of disciplines in the humanities and social sciences and in science and technology.

Catalogue
of the Type Fossils
in the Woodwardian
Museum, Cambridge

HENRY WOODS

CAMBRIDGE
UNIVERSITY PRESS

CAMBRIDGE UNIVERSITY PRESS

Cambridge, New York, Melbourne, Madrid, Cape Town, Singapore,
São Paolo, Delhi, Dubai, Tokyo, Mexico City

Published in the United States of America by Cambridge University Press, New York

www.cambridge.org
Information on this title: www.cambridge.org/9781108016032

This edition first published 1891
This digitally printed version 2010

ISBN 978-1-108-01603-2 Paperback

TYPE FOSSILS

IN THE

WOODWARDIAN MUSEUM.

LONDON: C. J. CLAY AND SONS,
CAMBRIDGE UNIVERSITY PRESS WAREHOUSE,
AVE MARIA LANE.

CAMBRIDGE: DEIGHTON, BELL, AND CO.
LEIPZIG: F. A. BROCKHAUS.
NEW YORK: MACMILLAN AND CO.

CATALOGUE OF THE TYPE FOSSILS

IN THE

WOODWARDIAN MUSEUM,

CAMBRIDGE.

BY

HENRY WOODS, B.A., F.G.S.

SCHOLAR OF ST JOHN'S COLLEGE.

WITH A PREFACE BY

T. McKENNY HUGHES, M.A., F.R.S.

WOODWARDIAN PROFESSOR OF GEOLOGY.

CAMBRIDGE:
AT THE UNIVERSITY PRESS.

1891

𝕮𝖆𝖒𝖇𝖗𝖎𝖉𝖌𝖊:

PRINTED BY C. J. CLAY, M.A., AND SONS,

AT THE UNIVERSITY PRESS.

PREFACE.

BY a 'type' is meant the original specimen to which any generic or specific name was first assigned. Subsequent observers in examining specimens which agree in general characters with an already described form, often notice differences which may indicate a new species, may be only due to incomplete description, or to the imperfect state of preservation of the type. In order to determine these points, it is necessary for them to see the actual fossil, which the author of the species had before him, when he wrote his description.

The importance of preserving and distinctly marking figured and described specimens, has only of late years been generally realised. A committee of the British Association reported upon the subject last year[1]. In the Woodwardian Museum such specimens have been mounted on tablets of a special colour,—at first pink was used, but now blue, a more stable colour is being substituted. The plan of exhibiting all the types by themselves, on the top of the cabinets was tried, and, except where they are mounted on coloured tablets, this method can be recommended, as in every museum of importance, inferior specimens are continually being replaced by better ones, and thus the type, which is sometimes a poor specimen may perhaps get lost sight of. As soon however as the types

[1] Rep. *Brit. Assoc.* (Leeds), 1890, p. 339.

were mounted on tablets of a conspicuous colour, we found that
they could be safely put into their proper places in the series,
and that it was better to display on the top of the cabinets
those specimens which best showed generic and specific characters, and were thus of greatest educational value and
general interest.

In addition to indicating in the museum the importance of
such specimens, it is most desirable to publish a catalogue of
them, so that specialists may know where to find the types, and
this task has been most ably performed for the Woodwardian
Museum by Mr H. Woods, whose knowledge of the Museum
and of Palæontology, eminently qualified him for the work.

Besides the types and other figured fossils, there are a great
number of specimens *referred to.* An author mentions for
instance, that there is in the Woodwardian Museum, in such
and such a series, a fossil which illustrates some point under
discussion. These specimens have been labelled as ' mentioned.'

There is another series, the acquired importance of which is
often almost equal to that of types, namely, the specimens
which have been determined for us by high authorities, and we
have endeavoured to indicate this in the case of all those which
we have recently acquired, or which have been lately determined, but it will be long before we can overtake the work of
indicating by labels who is responsible for the determination of
each specimen, even of those for which we know or can ascer-
tain the authority. But in a museum of any antiquity, the
authorities for the determination of the larger number of the
specimens must remain for ever unrecorded.

We have not included, in this catalogue, the described and
figured fossils in the collections of the seventeenth century, now
in our Museum. For instance, incorporated into the original
museum of Dr John Woodward, we have the collection of Agos-
tino Scilla, a distinguished painter and naturalist, who was born
at Messina in 1639. He subsequently removed to Rome, where
he became President of the Academy of Painting. A few of
his pictures are to be found in Rome, and the churches of

Messina possess a considerable number. He is mentioned in many places in the works of Boccone the great naturalist with whom he travelled in Sicily. Scilla published in 1670 a book[1] the object of which was to prove by direct comparison, that fossils were not, as many held in those days, merely accidental forms or freaks of nature, but really the remains of organisms which had lived, died, and been buried in the mud and sand. With a view to proving this, he collected fossil and recent shells, bones, etc., and made excellent pencil sketches of some of the specimens in his collection, from which the figures in his work were drawn. The collection itself was acquired by Dr Woodward, and the University now possesses, not only the very specimens upon which Scilla based his observations, but also the original drawings which he made for the engraver. Dr Woodward highly prized these drawings for he wrote on the fly leaf of the copy of the work with which they were bound up

"Liber ingentis Pretii quippe qui exhibet Archetypas Fossilium Imagines, ipsius Augustini Scillæ, præclari Pictoris, primo Messanæ, exinde Romæ, insigni Penicillo delineatas."

This book, with the original drawings, I found behind a case in the Woodwardian Museum, its very existence having been previously unknown.

Our own countryman Lister did not share the views of the Sicilian, and in 1688 published a work containing an article 'De Conchitis sive Lapidibus qui quandam similitudinem cum conchis marinis habeant.' We have the original fine specimen of *Productus giganteus*, which he looked upon as only a piece of stone having an accidental resemblance to a marine shell.

In the works of Dr Woodward very few fossils are figured, though many are described, and points of interest and of controversy in his time are referred to and illustrated by the specimens in his cabinets. For example, I remember the late

---

[1] *La Vana speculazione disingannata dal senso*, etc., Naples 1670. A Latin translation was published in Rome in 1747, and later editions in 1752 and 1759.

Dr Lightfoot coming to ask if it were possible to verify the statement of Woodward, that what had been taken for the bones of Christians massacred by the Turks at their taking of Philadelphia in Lydia consisted in reality " of various Bodies, chiefly Vegetables, incrusted over and cemented together by sparry and stoney Matter." The specimen was easily found by reference to Woodward's Catalogue, and the question answered.

These old collections contain specimens of great historical and geological interest, of which we hope by and bye to give a fuller account.

T. McKENNY HUGHES.

CONTENTS.

INTRODUCTION.

THIS Catalogue includes in addition to the types proper, those specimens which have been figured and described as examples of species previously defined.

The names at present accepted are printed in Clarendon type, the synonyms in italics. After a considerable portion of the work had been printed, it was suggested to me that it would have been better to have given only the names under which the specimens were described, or at any rate to have given greater prominence to these rather than to the names now accepted. In some ways this method would certainly have been more convenient than the one adopted.

In order to make the work more complete, I have gone through a large amount of palæontological literature, and have in this way found in the Museum many types whose existence here was previously unsuspected. Nevertheless, there can, I think, be no doubt that others will in time be discovered.

The following is a list of the authors who have described specimens in the Museum :—A. L. Adams, L. Agassiz, D. T. Ansted, L. Barrett, F. A. Bather, W. Bean, T. Bell, G. T. Bettany, J. F. Blake, C. J. F. Bunbury, P. H. Carpenter, W. Carruthers, J. Carter, C. Coignou, T. Davidson, J. W. Davis, J. Donald, P. M. Duncan, H. M. Edwards, R. Etheridge, R. Etheridge, jun., H. Falconer, T. W. Fletcher, A. H. Foord,

E. Forbes, J. S. Gardner, J. G. Goodchild, J. W. Gregory,
T. T. Groom, A. C. L. G. Günther, J. Haime, H. Hicks
G. J. Hinde, J. Hopkinson, W. H. Hudleston, T. H. Huxley,
H. Jelly, T. R. Jones, A. J. Jukes-Browne, W. Keeping,
J. W. Kirkby, W. Kowalevsky, P. Lake, E. R. Lankester,
C. Lapworth, J. Leckenby, W. Lonsdale, P. de Loriol, J. Lycett,
R. Lydekker, J. E. Marr, F. McCoy, L. C. Miall, J. Morris,
R. I. Murchison, E. T. Newton, R. Owen, J. Parkinson, A.
Pavlow, J. Phillips, J. Postlethwaite, T. Roberts, J. W. Salter,
R. Schäfer, H. G. Seeley, A. C. Seward, G. W. Shrubsole,
W. J. Sollas, J. Sowerby, J. de C. Sowerby, R. H. Traquair,
A. Wanklyn, G. F. Whidborne, S. V. Wood, A. S. Woodward,
H. Woodward, S. P. Woodward, T. Wright, J. Young.

In the preparation of the Vertebrate portion I have derived
very considerable help from the excellent 'Catalogue of British
Fossil Vertebrata' by Messrs. A. S. Woodward and C. D.
Sherborn.

I have much pleasure in expressing my thanks to several
gentlemen for the assistance they have very generously given
me. I would especially mention Mr A. C. Seward who has
revised the Plants, Dr G. J. Hinde the Incertæ sedis and
Porifera, Mr J. E. Marr the Hydrozoa and Trilobita, Professor
H. A. Nicholson the Actinozoa, Mr J. W. Gregory the Echi-
noidea, Asteroidea, and Blastoidea, Mr F. A. Bather the Cri-
noidea and Cystidea, Mr G. R. Vine the Polyzoa, Mr J. F.
Walker the Brachiopoda, Professor T. Rupert Jones the Ostra-
coda and Phyllocarida, Mr J. Carter the Decapoda, Mr A. H.
Foord the Cephalopoda, and Mr A. S. Woodward the Vertebrata.
I am also indebted to Mr T. Roberts and Mr J. W. Gregory for
much general assistance and advice.

H. WOODS.

November, 1891.

LIST OF THE CHIEF COLLECTIONS IN THE
WOODWARDIAN MUSEUM.

Aitken Collection.—This consists of Plants, Mollusca, and Fishes from the Coal Measures, and Mollusca from the Millstone Grit. It was purchased from Mr J. Aitken in 1877.

Barrande Collection.—Fossils from the Palæozoic formations of Bohemia. Purchased from M. Barrande in 1856.

Burrows Collection.—Fossils from the Carboniferous Limestone of Settle, Yorkshire, collected by the late Mr J. H. Burrows, from whom they were purchased in 1872. The collection contains specimens of Chitonidæ figured by Kirkby, and Brachiopoda by Davidson.

De Stefani Collection.—A collection of Italian Pliocene fossils purchased from Mr C. de Stefani in 1882.

Dover Collection.—This collection was formed by the late Mr W. Kinsey Dover, F.G.S., of Keswick, by whom it was presented in 1890. It consists chiefly of Trilobites and Graptolites from the Skiddaw Slates, and contains specimens figured by Etheridge, Nicholson, Hopkinson, Goodchild and Postlethwaite.

Fisher Collection.—The collection of the Rev. O. Fisher, M.A., consisting of (i) fossils from the Bracklesham and Barton beds, presented in 1875, (ii) Insects from the Purbeck beds of Durlstone Bay and Ridgway, presented in 1854, and (iii) sundry other fossils presented at various times. Some of the Eocene Mollusca have been figured by S. V. Wood, and the Purbeck Mollusca by De Loriol.

Fletcher Collection.—A large collection from the Wenlock Limestone, chiefly of Dudley, purchased from the late Capt. T. W. Fletcher. It contains several of Salter's types, some Corals figured by Edward and Haime, and a few Trilobites by Fletcher.

Forbes-Young Collection.—A series of fossils from the Chalk of Kent and Sussex, collected by the late Dr Forbes-Young, and presented by Sir Charles Young and Mr Henry Young in 1862.

Goodman Collection.—This was formed by the late Mr Neville Goodman, M.A., of St Peter's College, and was presented by his son Dr R. N. Goodman in 1890. It consists chiefly of fossils from the British and Foreign Tertiaries, with a few from the Mesozoic formations.

Hawkins Collection.—Saurians from the Lias, presented by Mr T. Hawkins in 1856.

Image Collection.—A collection of Cretaceous fossils, purchased from the late Rev. T. Image.

Leckenby Collection.—This large collection was formed by the late Mr John Leckenby of Scarborough, from whom it was purchased in 1871. It contains fossils from various formations, but is especially rich in those from the Jurassic and Cretaceous beds of Yorkshire, many of which have been figured, amongst these we may mention, the Ammonites from the Lias by Wright, the Plants from the Lower Oolites by Leckenby, the Jurassic Mollusca by Morris and Lycett, the Gasteropoda by Hudleston, and the Mollusca from the Kellaways Rock by Leckenby.

Monk Collection.—Fossils, chiefly Mollusca and Brachiopoda, from the Inferior Oolite of Somerset and Dorset, purchased from Mr H. Monk, of Yeovil, in 1885.

Montagu Smith Collection.—This was formed by the late Mr Montagu Smith, B.A., of Trinity College, and presented in 1883. It consists of fossils from the Pliocene deposits of East Anglia, from the Chalk Rock of Cuckhamsley, the Lower Chalk of Pyrton and Folkestone, the Gault of Folkestone, the Lower Greensand of Faringdon, and the Upper Jurassic formations.

Münster Collection.—This large collection of foreign fossils was purchased from the late Count Münster in 1840. It contains specimens from various formations, but is especially rich in Triassic and Jurassic forms.

Porter Collection.—Saurians from the Oxford Clay, forming part of the collection of the late Dr H. Porter, of Peterborough, purchased in 1866.

Strickland Collection.—A large and varied collection, formed by the late Mr H. E. Strickland, M.A., F.R.S. Bequeathed by Mrs Strickland 1888. It contains a few figured specimens.

Walton Collection.—The large collection of the late Mr W. Walton, of Bath, presented in 1876. It is especially rich in fossils from the Jurassic beds of the South of England, and contains Corals from the Great Oolite and Fuller's Earth figured by Edwards and Haime, Polyzoa from the Great Oolite and Bradford Clay by Haime, and Mollusca from the Great Oolite and Forest Marble by Lycett.

ABBREVIATIONS.

A. M. N. H.   Annals and Magazine of Natural History.

G. M.   Geological Magazine.

Q. J. G. S.   Quarterly Journal of the Geological Society of London.

Salter, Cat.   Catalogue of the Cambrian and Silurian Fossils in the Geological Museum of the University of Cambridge, by J. W. Salter. 1873.

ADDENDA ET CORRIGENDA.

Page 31. *Add* **Echinoconus subrotundus** (Mantell), T. Roberts, G. M. dec. 3, VIII (1891), p. 116, woodcuts. *Upper Chalk; Shudy Camps.* Presented by J. Carter, Esq., F.G.S.

„ 33. *Add* **Peltastes wrighti** (Desor), T. Roberts, G. M. dec. 3, VIII (1891), p. 118, woodcuts 1. *Lower Greensand; Faringdon.* Montagu Smith Collection.

„ 34. *Add* **Schizaster cuneatus**, J. W. Gregory, Proc. Geol. Assoc. XII (1891), pl. i, f. 1, 2, 3, p. 24. *London Clay; Bognor.*

„ 37. *Add* **HOLOTHUROIDEA.**

Chirodota convexa, G. F. Whidborne, Q. J. G. S. XXXIX (1883), pl. xix, f. 14, 14*a*, p. 537. *Inferior Oolite; Burton Bradstock.* Presented by the Rev. G. F. Whidborne, M.A.

Chirodota? **gracillima**, G. F. Whidborne, Q. J. G. S. XXXIX (1883), pl. xix, f. 15, 15*a*, p. 537. *Inferior Oolite; Burton Bradstock.* Presented by the Rev. G. F. Whidborne, M.A.

„ 39, lines 7 and 24, *dele* The specimen is missing.

„ 45, line 10 from Polyzoa, *for* 1855 *read* 1851.

„ 47, line 16, *for* 1855 *read* 1851.

„ 48, lines 26 and 33, *for* 1855 *read* 1851.

„ 129. *Add* **Orthoceras cf. elongatocinctum**, Portlock, A. H. Foord, Q. J. G. S. XLVII (1891), p. 526. *Keisley Limestone; Keisley.*

„ 129. *Add* **Orthoceras pusgillense**, A. H. Foord, Q. J. G. S. XLVII (1891), p. 527, woodcuts. *Corona-beds; Pusgill.* Presented by Professor H. A. Nicholson, M.D.

„ 130. *Add* **Orthoceras sp.**, A. H. Foord, Q. J. G. S. XLVII (1891), p. 526. *Staurocephalus Limestone; Swindale.* Presented by J. E. Marr, Esq., F.R.S.

„ 131. *Add* **Pleuronautilus nodoso-carinatus** (Römer), A. H. Foord, G. M. dec. 3, VIII (1891), p. 481, woodcut *a.* *Millstone Grit; Caton, Lancashire.* Presented by A. C. Seward, Esq., M.A.

The dates given for the Reports of the British Association are those of the years for which the volumes are issued, the actual date of publication being the year following. The date of J. W. Salter's Appendix A to McCoy's British Palæozoic Fossils should have been given as 1852 instead of 1855.

CATALOGUE OF TYPE FOSSILS IN THE WOODWARDIAN MUSEUM.

INCERTÆ SEDIS.

Astylospongia grata, Salter, Cat. (1873), p. 40. G. J. Hinde, Brit. Foss. Sponges (1888), p. 177. *Coniston Limestone; Coniston Water.*

Chondrites acutangulus, F. McCoy, Q. J. G. S. IV (1848), p. 224, and Brit. Palæoz. Foss. (1851), pl. i A, f. 5. *Skiddaw Slates; Low Fell, Loweswater.*

C. infòrmis, F. McCoy, Q. J. G. S. IV (1848), p. 223, and Brit. Palæoz. Foss. (1851), pl. i A, f. 4. *Skiddaw Slates; Whitless.*

C. verisimilis, Salter, Cat. (1873), p. 99, woodcut. *Wenlock Limestone; Dudley.* Fletcher Collection.

C. ? sp., H. A. Nicholson, G. M. VI (1869), pl. xviii, f. D, p. 497. *Skiddaw Slates; Rake Beck, near Melmerby.* Presented by Prof. H. A. Nicholson.

Eophyton ? palmatum, H. A. Nicholson, G. M. VI (1869), pl. xviii, f. C, p. 497. *Skiddaw Slates; Barf, near Keswick.* Presented by Prof. H. A. Nicholson.

Ischadites ? micropora, Salter, Cat. (1873), p. 40. G. J. Hinde, Brit. Foss. Sponges (1888), p. 179. *Middle Bala; Blaen-y-cwm, Llansaintffraid.*

Myrianites tenuis, F. McCoy, A. M. N. H. ser. 2, VII (1851), p. 394, and Contrib. Brit. Pal. (1854), p. 178, and Brit. Palæoz. Foss. (1851), pl. i D, f. 13, p. 130. *Bala Beds; Grieston.*

Another specimen, H. A. Nicholson and R. Etheridge, jun.,

Mon. Sil. Foss. Girvan, I, fascic. III (1880), p. 312, woodcut 10. *Gala Group; Thornilee, Peebleshire.* Presented by Professor H. A. Nicholson, D.Sc.

Pasceolus goughi, Salter, Cat. (1873), p. 175. Undetermined fossil, F. McCoy, Brit. Palæoz. Foss. (1851), pl. i D, f. 9, 9 *a*. *Upper Ludlow; Benson Knot.*

Protovirgularia dichotoma, F. McCoy, A. M. N. H. ser. 2, VI (1850), p. 272, and Contrib. Brit. Pal. (1854), p. 156, and Brit. Palæoz. Foss. (1851), pl. i B, f. 11, 11 *a*, 12, 12 *a*, p. 10. *Bala Beds; near Moffat.*

Sphærospongia hospitalis, Salter, Cat. (1873), p. 40. G. J. Hinde, Q. J. G. S. XL (1884), p. 835, and Brit. Foss. Sponges (1888), p. 182. *Bala Beds; Onny River.* This specimen is missing.

Spongarium æquistriatum, F. McCoy, A. M. N. H. ser. 2, VI (1850), p. 281, and Contrib. Brit. Pal. (1854), p. 167, and Brit. Palæoz. Foss. (1851), pl. i B, f. 15, 15 *a*, p. 42. *Upper Ludlow; Benson Knot, Kendal.*

S. interlineatum, F. McCoy, A. M. N. H. ser. 2, VI (1850), p. 281, and Contrib. Brit. Pal. (1854), p. 167, and Brit. Palæoz. Foss. (1851), pl. i B, f. 14, 14 *a*, p. 43. *Lower Ludlow; Benson Knot, Kendal.*

S. interruptum, F. McCoy, A. M. N. H. ser. 2, VI (1850), p. 282, and Contrib. Brit. Pal. (1854), p. 168, and Brit. Palæoz. Foss. (1851), pl. i B, f. 16, 16 *a*, 17, p. 43. *Upper Ludlow; Spital, Kendal.*

Stellascolites radiatus, R. Etheridge, in J. C. Ward, Geol. N. Part Eng. Lake District (1873), pl. xiii, p. 110. *Skiddaw Slates; Randal Crag, Skiddaw.* Dover Collection.

Verticillipora palmata [Salter, MS.], G. J. Hinde, Brit. Foss. Sponges (1888), p. 184. 'Sponge, new. Palmate lobed.' Salter, Cat. (1873), p. 100. *Wenlock Limestone; Dudley.* Fletcher Collection.

Vioa prisca, F. McCoy, Brit. Palæoz. Foss. (1852), pl. i B, f. 1, 1 *a*, p. 260. Salter, Cat. (1873), p. 85. G. J. Hinde, Brit. Foss. Sponges (1888), p. 184. *May Hill Group; Malvern.*

PLANTÆ.

Anomozamites minus (Brongniart). *Pterophyllum minus,* J. Leckenby, Q. J. G. S. xx (1864), pl. ix, f. 2, p. 78. *Inferior Oolite; near Scarborough.* Leckenby Collection.

Araucarites phillipsi, W. Carruthers, G. M. vi (1869), pl. ii, f. 8, 9, p. 6. *Inferior Oolite; Yorkshire.* Leckenby Collection.

Butrotrephis harknessi, v. Protannularia harknessi.

Calamites undulatus, Sternberg, A. C. Seward, G. M. dec. 3, v (1888), pl. ix, f. 1 A, 1 B, p. 289. *Coal Measures; Wigan.*

C. sp., A. C. Seward, G. M. dec. 3, v (1888), pl. ix, f. 2 A, 2 B, p. 290. *Coal Measures; locality unknown.*

Ctenis leckenbyi, v. Ptilozamites leckenbyi.

Cycadites zamioides, J. Leckenby, Q. J. G. S. xx (1864), pl. xi, f. 1, p. 77. J. Phillips, Geol. Yorkshire, part i, edition 3 (1875), p. 228, woodcut 58. *Inferior Oolite (Upper Shale); Scarborough.* Leckenby Collection.

Cycadostrobus elegans, Carruthers, J. S. Gardner, Rep. Brit. Assoc. (1886), pl. vii, f. 8, p. 244. *Wealden; Brook Point, Isle of Wight.*

Cyclopteris sp., A. C. Seward, G. M. dec. 3, v (1888), pl. x, p. 344. *Upper Coal Measures; Brierly Common, Yorkshire.*

Dicksonia arguta (Lindley and Hutton). *Neuropteris arguta,* J. Leckenby, Q. J. G. S. xx (1864), pl. x, f. 4, p. 79. *Inferior Oolite; near Scarborough.* Leckenby Collection.

D. nephrocarpa (Bunbury). *Sphenopteris nephrocarpa,* C. J. F. Bunbury, Q. J. G. S. vii (1851), pl. xii, f. 1 a, b, p. 179. *Inferior Oolite; near Scarborough.* Leckenby Collection.

1—2

Fucoides erectus [Bean MS.], J. Leckenby, Q. J. G. S. xx (1864), pl. xi, f. 3 *a*, *b*, p. 81. J. Phillips, Geol. Yorkshire, part i, edition 3 (1875), p. 196, woodcut 3. *Inferior Oolite; Cloughton Wyke.* Leckenby Collection.

Gleichenia hantoniensis (Wanklyn). J. S. Gardner and C. Ettingshausen, Brit. Eoc. Flora, I (1880), pl. x, f. 2—4, pp. 43, 59. *Mertensites hantoniensis* and *M. crenata*, A. Wanklyn, A. M. N. H. ser. 4, III (1869), pl. i, f. 1 *b*, *c*, *d*, 2, 3, pp. 11, 12. *Middle Bagshot Beds; Bournemouth.*

Kaidacarpum minus, W. Carruthers, G. M. v (1868), p. 156. J. S. Gardner, G. M. dec. 3, III (1886), p. 496. *Lower Greensand; Potton.*

Matonidium göpperti (Schimper). *Pecopteris polydactyla*, Goeppert, J. Leckenby, Q. J. G. S. xx (1864), pl. xi, f. 1 *a*, *b*, p. 80. *Inferior Oolite (Lower Shale); Scarborough.* Leckenby Collection.

Mertensites crenata, v. Gleichenia hantoniensis.

M. hantoniensis, v. Gleichenia hantoniensis.

Microdictyon woodwardi (Leckenby). *Phlebopteris woodwardi*, J. Leckenby, Q. J. G. S. xx (1864), pl. viii, f. 6, p. 81. *Inferior Oolite; near Scarborough.* Leckenby Collection.

Neuropteris arguta, v. Dicksonia arguta.

Nilssonia compta (Lindley and Hutton). *Pterophyllum comptum*, J. Leckenby, Q. J. G. S. xx (1864), pl. ix, f. 1, p. 77. *Inferior Oolite (Upper Shale); Scarborough.* Leckenby Collection.

Odontopteris leckenbyi, v. Ptilozamites leckenbyi.

Osmunda lignitum (Giebel). *Cyathæ?* A. Wanklyn, A. M. N. H. ser. 4, III (1869), pl. i, f. 5. *M. Bagshot Beds; Bournemouth.*

Otopteris graphica [Bean MS.], J. Leckenby, Q. J. G. S. xx (1864), pl. viii, f. 5, p. 78. *Inferior Oolite; near Scarborough.* Leckenby Collection.

O. lanceolata, v. Otozamites gracilis.

O. mediana, v. Otozamites mediana.

O. tenuata, v. Otozamites bunburyanus.

Otozamites bunburyanus, Zigno. *Otopteris tenuata* [Bean MS.], J. Leckenby, Q. J. G. S. xx (1864), pl. ix, f. 3, p. 79. *Inferior Oolite; near Scarborough.* Leckenby Collection.

Otozamites gracilis (Phillips). *Otopteris lanceolata* [Bean MS.], J. Leckenby, Q. J. G. S. xx (1864), pl. viii, f. 4, p. 78. *Inferior Oolite; Scarborough.* Leckenby Collection.

O. gramineus, J. Phillips, Geol. Yorkshire, part i, edition 3 (1875), p. 223, woodcut 51. *Inferior Oolite; Whitby?* Leckenby Collection.

O. mediana (Leckenby). *Otopteris mediana*, J. Leckenby, Q. J. G. S. xx (1864), pl. x, f. 2, p. 78. *Inferior Oolite (Lower Shale); Scarborough.* Leckenby Collection.

Palæozamia pecten (Lindley and Hutton). J. Leckenby, Q. J. G. S. xx (1864), pl. ix, f. 4, p. 77. *Inferior Oolite; Scarborough.* Leckenby Collection.

Pecopteris insignis, J. Lindley and W. Hutton, Foss. Flora, II (1833—5), pl. cvi, p. 69. *Inferior Oolite; Gristhorpe.* (Counterpart of the Type.) Leckenby Collection.

P. polydactyla v. Matonidium göpperti.

Phlebopteris woodwardi, v. Microdictyon woodwardi.

Pinites carruthersi, J. S. Gardner, G. M. dec. 3, III (1886), p. 498, and Rep. Brit. Assoc. (1886), pl. vii, f. 6, p. 244. *Wealden; Brook Point, Isle of Wight.*

P. cylindroides, J. S. Gardner, G. M. dec. 3, III (1886), p. 499, and Rep. Brit. Assoc. (1886), pl. vii, f. 2, 2 *a*, p. 245. *Lower Greensand; Potton.*

P. leckenbyi, W. Carruthers, G. M. VI (1869), pl. i, f. 1—5, p. 2. J. H. Balfour, Palæontological Botany (1872), pl. ii, f. 4. *Wealden; Brook, Isle of Wight.* Leckenby Collection.

P. pottonensis, J. S. Gardner, G. M. dec. 3, III (1886), p. 499, and Rep. Brit. Assoc. (1886), pl. vii, f. 3, p. 245. *Lower Greensand; Potton.*

P. valdensis, J. S. Gardner, G. M. dec. 3, III (1886), p. 498, and Rep. Brit. Assoc. (1886), pl. vii, f. 4, p. 244. *Wealden; Brook Point, Isle of Wight.*

P. sp., J. S. Gardner, G. M. dec. 3, III (1886), p. 499, and Rep. Brit. Assoc. (1886), pl. vii, f. 7, p. 245. *Wealden; Brook Point.*

Pinus dixoni (Bowerbank), J. S. Gardner, Brit. Eoc. Flora, II (1884), pl. xiii, f. 1, 2, 8, p. 66. *Barton Beds; Highcliff.* Fisher

Collection. (Not from Bracklesham as stated by Gardner in the work mentioned.) One specimen (fig. 8) has perished.

Pterophyllum angustifolium [Bean MS.], J. Leckenby, Q. J. G. S. xx (1864), pl. viii, f. 2, p. 77. J. Phillips, Geol. Yorkshire, part i, edition 3 (1875), p. 227, woodcut 56. *Inferior Oolite; Gristhorpe.* Leckenby Collection.

P. comptum, v. Nilssonia compta.

P. minus, v. Anomozamites minus.

P. medianum [Bean MS. sp.], J. Leckenby, Q. J. G. S. xx (1864), pl. viii, f. 3, p. 77. J. Phillips, Geol. Yorkshire, part i, edition 3 (1875), p. 226, woodcut 55. *Inferior Oolite; Scarborough.* Leckenby Collection.

Ptilozamites leckenbyi [Bean MS. sp.]. *Ctenis leckenbyi*, J. Leckenby, Q. J. G. S. xx (1864), pl. x, f. 1 *a, b,* p. 78. *Odontopteris leckenbyi*, J. Phillips, Geol. Yorkshire, part i, edition 3 (1875), p. 218, woodcut 41. *Inferior Oolite (Upper Shale); Yorkshire.* Leckenby Collection.

Protannularia harknessi (Nicholson). *Butrotrephis harknessi,* H. A. Nicholson, G. M. vi (1869), pl. xviii, f. A, B, p. 495. *Skiddaw Slates; Thornship Beck, Shap.* Presented by Prof. H. A. Nicholson, D.Sc.

Phlebopteris woodwardi, v. Microdictyon woodwardi.

Sphenopteris göpperti, Dunker. *Sphenopteris jugleri*, Ettingshausen, J. Leckenby, Q. J. G. S. xx (1864), p. 79. J. Phillips, Geol. Yorkshire, part i, edition 3, p. 218, woodcut 40. *Inferior Oolite (Lower Shale); Scarborough.* Leckenby Collection.

S. jugleri, v. Sphenopteris göpperti.

S. modesta [Bean MS.], J. Leckenby, Q. J. G. S. xx (1864), pl. x, f. 3 *a, b,* p. 79. *Inferior Oolite (Lower Shale); Scarborough.* Leckenby Collection.

S. nephrocarpa, v. Dicksonia nephrocarpa.

Thyrsopteris racemosa (Lindley and Hutton). *Tympanophora simplex et racemosa*, J. Leckenby, Q. J. G. S. xx (1864), pl. xi, f. 2, p. 79. *Inferior Oolite; Scarborough.* Leckenby Collection.

Tympanophora simplex, v. Thyrsopteris racemosa.

'Fruits.' W. Keeping, Foss. Upware and Brickhill (1883), pl. viii, f. 7, 7 *a*, p. 150. *Lower Greensand; Upware.*

ANIMALIA.

PORIFERA.

Astylospongia grata, see under Incertæ sedis.

Chenendopora expansa (Benett), *var.*, H. G. Seeley, A. M. N. H. ser. 3, XVII (1866), p. 181. *Red Chalk; Hunstanton.*

Corynella nodosa, W. Keeping, Foss. Upware and Brickhill (1883), pl. viii, f. 4, p. 148. *Lower Greensand; Brickhill.*

C. sp., W. Keeping, Foss. Upware and Brickhill (1883), pl. viii, f. 5, p. 149. *Lower Greensand; Brickhill.*

Dictyophyton danbyi (McCoy), G. J. Hinde, Cat. Foss. Sponges, Brit. Mus. (1883), p. 131, and Brit. Foss. Sponges (1888), pl. ii, f. 4, 4 *a*, 4 *c*, p. 128. *Tetragonis danbyi*, F. McCoy, Brit. Palæoz. Foss. (1851), pl. i D, f. 7, 7 *a*, 8, p. 62. *Upper Ludlow; Benson Knot and Brigsteer, Kendal.*

Eubrochus clausus, W. J. Sollas, G. M. dec. 2, III (1876), pl. xiv, f. 1, p. 398. G. J. Hinde, Cat. Foss. Sponges, Brit. Mus. (1883), p. 129. *Cambridge Greensand; Cambridge.*

Hallirhoa costata (Lamouroux). *Siphonia costata*, W. J. Sollas, Q. J. G. S. XXXIII (1877), pl. xxv, f. 2, 2 *a*, p. 811. *Upper Greensand; Wiltshire.*

Hyalostelia fasciculus (McCoy), G. J. Hinde, Brit. Foss. Sponges (1888), p. 111. *Pyritonema fasciculus*, F. McCoy, A. M. N. H. ser. 2, VI (1850), p. 273, and Contrib. Brit. Pal. (1854), p. 157, and Brit. Palæoz. Foss. (1851), pl. i B, f. 13, 13 *a, b*, p. 10. *Llandeilo Beds; Tre-gib, near Llandeilo.*

Ischadites kœnigi, Murchison, J. W. Salter, Cat. (1873), p. 100.
G. J. Hinde, Q. J. G. S. XL (1884), p. 801, and Brit. Foss.
Sponges (1881), pl. ii, f. 1, p. 120. *Wenlock Limestone; Dudley.*
Fletcher Collection.

I.? micropora, see under Incertæ sedis.

Jerea pastinacea, F. McCoy, A. M. N. H. ser. 2, II (1848),
p. 398, and Contrib. Brit. Pal. (1854), p. 45. *Upper Greensand; Vale of Pewsey, Wiltshire.* This specimen cannot be
identified with certainty.

J. websteri (Sowerby?). *Siphonia websteri*, W. J. Sollas, Q. J. G.
S. XXXIII (1877), pl. xxv, f. 7, 7 *a*, p. 806. *Gault; Folkestone.*

Manon foliaceum, F. McCoy, A. M. N. H. ser. 2, II (1848),
p. 399, and Contrib. Brit. Pal. (1854), p. 46. *Chalk; probably
Flamborough.* (Not from the Coralline Oolite of Malton as
stated by McCoy.)

M. reussi, v. Verruculina reussi.

Ophrystoma ocellatum [Seeley MS. sp.], G. J. Hinde, Cat. Foss.
Sponges, Brit. Mus. (1883), p. 126. *Porospongia*, O. Fisher,
Q. J. G. S. XXIX (1873), pl. vi, f. 1. *Ventriculites cavatus*,
Smith, W. J. Sollas, *ibid.* (1873), p. 68. *Cambridge Greensand;
Cambridge.*

Pachytilodia sp., W. Keeping, Foss. Upware and Brickhill (1883),
pl. viii, f. 6, p. 149. *Lower Greensand; Upware.*

Pharetrospongia strahani, W. J. Sollas, Q. J. G. S. XXXIII
(1877), pl. xi, p. 242. *Cambridge Greensand; Cambridge.*

Plocoscyphia laxa, F. McCoy, A. M. N. H. ser. 2, II (1848),
p. 397, and Contrib. Brit. Pal. (1854), p. 45. *Upper Greensand;
Lyme Regis.*

P. pertusa, Geinitz, W. Keeping, Foss. Upware and Brickhill
(1883), pl. viii, f. 1 *a, b*, p. 145. *Lower Greensand; Brickhill.*

Protospongia diffusa, J. W. Salter, Rep. Brit. Assoc. (1865),
p. 285, and Cat. (1873), p. 3. G. J. Hinde, Brit. Foss. Sponges
(1888), p. 180. *Menevian Group; Pen-pleidau, St Davids.*

P. fenestrata, v. Protospongia hicksi.

Protospongia hicksi, G. J. Hinde, Brit. Foss. Sponges (1888), pl. i, f. 2, 2 *a*, p. 107. *P. fenestrata*, H. Hicks (*non* Salter), Q. J. G. S. XXVII (1871) pl. xvi, f. 20, p. 401. W. J. Sollas, *ibid.* XXXVI (1880), p. 363, woodcut. *Menevian Group; Porth-y-rhaw, St Davids.*

P.? major, H. Hicks, Q. J. G. S. XXVII (1871), pl. xvi, f. 14—17, p. 401. G. J. Hinde, Brit. Foss. Sponges (1888), p. 180. *Harlech Group; St Davids.*

Pyritonema fasciculus, v. Hyalostelia fasciculus.

Scyphia tessellata, H. G. Seeley, A. M. N. H. ser. 3, XVII (1866), p. 181. *Red Chalk; Hunstanton.*

Sphœrospongia hospitalis, see under Incertæ sedis.

Siphonia costata, v. Hallirhoa costata.

S. websteri, v. Jerea websteri.

Stauronema carteri, W. J. Sollas, A. M. N. H. ser. 4, XIX (1877), pl. i, f. 1—4, p. 23. *Upper Greensand; Folkestone.*

S. lobata, W. J. Sollas, A. M. N. H. ser. 4, XIX (1877), pl. i, f. 5—7, p. 24. *Upper Greensand; Folkestone.*

Tetragonis danbyi, v. Dictyophyton danbyi.

Tremacystia clavata (Keeping). *Verticellites clavata*, W. Keeping, Foss. Upware and Brickhill (1883), pl. viii, f. 3, p. 146. *Lower Greensand; Upware.*

Ventriculites cavatus, v. Ophrystoma ocellatum.

Verruculina reussi (McCoy). *Manon reussi*, F. McCoy, A. M. N. H. ser. 2, II (1848), p. 398, and Contrib. Brit. Pal. (1854), p. 46. *Upper Chalk; Flamborough.*

Verticellites annulatus, W. Keeping, Foss. Upware and Brickhill (1883), pl. viii, f. 2, p. 146. *Lower Greensand; Upware.*

V. clavata, v. Tremacystia clavata.

Verticillipora palmata, see under Incertæ sedis.

Vioa prisca, see under Incertæ sedis.

CŒLENTERATA.

HYDROZOA.

Callograptus radiatus, J. Hopkinson, Rep. Brit. Assoc. (1872), p. 108, and Q. J. G. S. xxxi (1875), pl. xxxvi, f. 8 *a, b*, p. 665. *Lower Arenig. Fig.* 8 *a, from north of Trwyn-hwrddyn, White-sand Bay.* Presented by J. Hopkinson, Esq., F.G.S. *Fig.* 8 *b, from Road-uchaf, Ramsey Island.* Presented by D. Homfray, Esq.

C. salteri, Hall, J. Hopkinson, Q. J. G. S. xxxi (1875), pl. xxxvi, f. 10, p. 667. *Middle Arenig; Whitesand Bay.* Presented by J. Hopkinson, Esq. F.G.S.

Climacograptus scalaris, Hall. *Diplograptus rectangularis*, F. McCoy, A. M. N. H. ser. 2, vi (1850), p. 271, and Contrib. Brit. Pal. (1854), p. 155, and Brit. Palæoz. Foss. (1851), pl. i B, f. 8—10, p. 8. *Birkhill Shales; Moffat.*

Dendrograptus arbuscula, Salter, Cat. (1873), p. 21. *Middle Arenig; Whitesand Bay.* Presented by Dr H. Hicks, F.R.S.
Other specimens, J. Hopkinson, Q. J. G. S. xxxi (1875), pl. xxxvi, f. 5 *a, b*, p. 663. *Lower Arenig; fig.* 5*a from Whitesand Bay, f.* 5 *b, from Road-uchaf, Ramsey Island.* Presented by Dr H. Hicks.

D. persculptus, J. Hopkinson, Q. J. G. S. xxxi (1875), pl. xxxvi, f. 4 *a—d*, p. 663. *Lower Arenig; creek north of Trwyn-hwrddyn, Whitesand Bay.* Presented by Dr H. Hicks, F.R.S.

D. ramsayi, J. Hopkinson, Q. J. G. S. xxxi (1875), pl. xxxvii, f. 2 *a*, 2 *b*, p. 664. *Lower Llandeilo; Abereiddy Bay.* Presented by Dr H. Hicks.

D. serpens, J. Hopkinson, Q. J. G. S. xxxi (1875), pl. xxxvii, f. 3, p. 665. *Lower Llandeilo; Abereiddy Bay.* Presented by Dr H. Hicks.

Dictyograptus (Desmograptus) cancellatus, J. Hopkinson, Q. J. G. S. xxxi (1875), pl. xxxvi, f. 11, p. 668. *Lower Arenig; north of Trwyn-hwrddyn, Whitesand Bay.* (Counterpart of the type.) Presented by J. Hopkinson, Esq. F.G.S.

Dictyograptus homfrayi, J. Hopkinson, Q. J. G. S. xxxi (1875), pl. xxxvi, f. 13, p. 668. *Lower Arenig; Road-uchaf, Ramsey Island.* Presented by D. Homfray, Esq.

Didymograptus bifidus (Hall), J. Hopkinson, Q. J. G. S. xxxi (1875), pl. xxxiii, f. 8 *d*, p. 646. *Upper Arenig; Porth-hayog, St Davids.* Presented by Dr H. Hicks.

D. euodus, C. Lapworth, Q. J. G. S. xxxi (1875), pl. xxxv, f. 1 *a—c*, p. 645. *Lower Llandeilo; Abereiddy Bay.* Presented by Dr H. Hicks.

D. extensus (Hall), J. Hopkinson, Q. J. G. S. xxxi (1875), pl. xxxiii, f. 1 *a—d*, p. 642. *Lower Arenig; Road-uchaf, Ramsey Island.* Presented by J. Hopkinson, Esq.
Another specimen, J. Hopkinson, A. M. N. H. ser. 5, ix (1882), p. 56, woodcuts 3, 3 *a*. *Skiddaw Slates; Randal Crag, Skiddaw.* Dover Collection.

D. furcillatus, C. Lapworth, Q. J. G. S. xxxi (1875), pl. xxxv, f. 3 *b, c*, p. 649. *Lower Llandeilo; Abereiddy Bay.* Presented by Dr H. Hicks. The specimen does not quite agree with the figure.

D. gibberulus, H. A. Nicholson, A. M. N. H. ser. 4, xvi (1875), pl. vii, f. 3, 3 *a, b*, p. 271. *Skiddaw Slates; fig. 3 from Randal Crag, fig. 3 a, b, from White Horse Fell, Skiddaw.* Dover Collection.

D. latus, Salter, Cat. (1873), p. 20. *Graptolites latus,* F. McCoy, Q. J. G. S. iv (1848), p. 223, woodcuts. Portions of various Dichograptid stipes, H. A. Nicholson, Q. J. G. S. xxiv (1868), p. 141. (Exclude the specimens described by McCoy from the Wenlock Shale, Builth, v. Monograptus priodon.) *Skiddaw Slates; Knockmurton.*

D. murchisoni (Beck), C. Lapworth, Q. J. G. S. xxxi (1875), pl. xxxv, f. 2 *a* [*b, f*, ?missing], p. 648. *Lower Llandeilo; Abereiddy Bay.* Presented by Dr H. Hicks.

D. pennatulus (Hall), J. Hopkinson, Q. J. G. S. xxxi (1875), pl. xxxiii, f. 3 *a, e*, p. 643. *Lower Arenig; Road-uchaf Ramsey Island.* Presented by J. Hopkinson, Esq. F.G.S.

Didymograptus sparsus, J. Hopkinson, Rep. Brit. Assoc. (1872), p. 108, and Q. J. G. S. xxxi (1875), pl. xxxiii, f. 2 *a—d*, p. 643. *Lower Arenig; Road-uchaf, Ramsey Island.* Presented by J. Hopkinson, Esq. F.G.S.

Diplograptus foliaceus (Murchison), C. Lapworth, Q. J. G. S. xxxi (1875), pl. xxxv, f. 7 *a, b*, p. 656. *Lower Llandeilo; Abereiddy Bay.* Fig. 7 *a*, presented by J. Hopkinson, Esq. Fig. 7 *b*, presented by Dr H. Hicks.

D. persculptus [Salter MS.], W. Carruthers, G. M. v (1868), p. 130. *Llandeilo Beds?; Gogofau, Carmarthenshire.*

D. rectangularis, v. Climacograptus scalaris.

Glossograptus ciliatus, Emmons, C. Lapworth, Q. J. G. S. xxxi (1875), pl. xxxiv, f. 7 *a—c*, p. 659. *Llanvirn Beds; Llanvirn Quarry.* Presented by Dr H. Hicks.

Graptolites latus, v. Didymograptus latus and Monograptus priodon.

G. lobiferus, v. Monograptus lobiferus.

G. millipeda, v. Monograptus lobiferus.

G. sedgwicki, v. Monograptus sedgwicki.

G. tenuis, v. Monograptus tenuis.

Monograptus lobiferus (McCoy). *Graptolites lobiferus*, F. McCoy, A. M. N. H. ser. 2, vi (1850), p. 270, and Contrib. Brit. Pal. (1854), p. 154, and Brit. Palæoz. Foss. (1855), pl. i B, f. 3, 3 *a*, p. 4. *Birkhill Shales; Moffat.*
    Graptolites millipeda, F. McCoy, A. M. N. H. ser. 2, vi (1850), p. 271, and Contrib. Brit. Pal. (1854), p. 155, and Brit. Palæoz. Foss. (1855), pl. i B, f. 6, 6 *a*, p. 5. *Birkhill Shales; Moffat.*

M. priodon (Bronn). *Graptolites latus*, F. McCoy (*pars*), Brit. Palæoz. Foss. (1851), pl. i B, f. 7, 7 *a—c*, p. 4. *Wenlock Shale; Builth Bridge.*

M. sedgwicki (Portlock). *Graptolites sedgwicki*, F. McCoy, Brit. Palæoz. Foss. (1851), pl. i B, f. 2, 2 *a*, p. 6. *Birkhill Shales; Moffat.*

M. tenuis (Portlock). *Graptolites tenuis*, F. McCoy, Brit. Palæoz. Foss. (1851), pl. i B, f. 4, 5, p. 6. *Birkhill Shales; Moffat.*

Nemagraptus capillaris, Emmons, C. Lapworth, Q. J. G. S. XXXI (1875), pl. xxxiv, f. 2 *a*, *b*, p. 653. *Llanvirn Beds; Llanvirn Quarry.*

Stachyodes verticillata (McCoy), H. A. Nicholson, Mon. Brit. Stromatop. (1886), p. 107. *Stromatopora (Caunopora) verticillata*, F. McCoy, A. M. N. H. ser. 2, VI (1850), p. 377, and Brit. Palæoz. Foss. (1851), p. 67, woodcuts *a* and *b*, on p. 66. *Devonian ; Teignmouth.*

Stromatopora (Caunopora) verticillata, v. Stachyodes verticillata.

Tetragraptus hicksi, J. Hopkinson, Q. J. G. S. XXXI (1875), pl. xxxiii, f. 12 *b*, p. 651. *Middle Arenig ; Whitesand Bay.* Presented by Dr H. Hicks.

T. serra (Brongniart), J. Hopkinson, A. M. N. H. ser. 5, IX (1882), p. 55, woodcuts 1, 2, 2 *a*. *Skiddaw Slates ; Randal Crag, Skiddaw.* Dover Collection.

Thamnograptus doveri, H. A. Nicholson, A. M. N. H. ser. 4, XVI (1875), pl. vii, f. 1, p. 271. *Skiddaw Slates; Randal Crag, Skiddaw.* Dover Collection.

ACTINOZOA.

Acervularia luxurians (Eichwald), H. M. Edwards and J. Haime, Brit. Foss. Corals (1854), pl. lxix, f. 2 *a*, *b*, p. 292. *Wenlock Limestone; Dudley.* Fletcher Collection.

Actinocystis grayi (Edwards and Haime). *Cystiphyllum grayi*, H. M. Edwards and J. Haime, Brit. Foss. Corals (1854), pl. lxxii, f. 3, 3 *a*, p. 297. *Wenlock Limestone; Dudley.* Fletcher Collection.

Alveolites fletcheri [Seeley MS.], Salter, Cat. (1873), p. 107. *Wenlock Limestone ; Dudley.* Fletcher Collection.

A. repens (Fougt), H. M. Edwards and J. Haime, Brit. Foss. Corals (1854), pl. lxii, f. 1, 1 *a*, p. 263. *Wenlock Limestone; Dudley.* Fletcher Collection.

A. seeleyi, Salter, Cat. (1873), p. 107. *Wenlock Limestone; Dudley.* Fletcher Collection.

A.? seriatoporoides, see under Polyzoa.

Alveolites vermicularis, F. McCoy, A. M. N. H. ser. 2, VI (1850), p. 377, and Brit. Palæoz. Foss. (1851), p. 69, woodcut. *Devonian; Bedruthen Steps, St Eval.*

Arachnophyllum typus, F. McCoy, A. M. N. H. ser. 2, VI (1850), p. 278, and Contrib. Brit. Pal. (1854), p. 163, and Brit. Palæoz. Foss. (1851), pl. i B, f. 27, 27 *a*, p. 38. *Wenlock Limestone; near Aymestry.*

Astræa (Palastræa) carbonaria, v. Cyathophyllum regium.

A. explanulata, v. Isastræa explanulata.

A. helianthella, v. Thamnastræa defranciana.

A. tenuistriata, v. Isastræa tenuistriata.

Aulacophyllum mitratum (Hisinger), H. M. Edwards and J. Haime, Brit. Foss. Corals (1854), pl. lxvi, f. 1 *a, b,* p. 280. *Wenlock Limestone; Dudley.* Fletcher Collection.

Axopora fisheri, P. M. Duncan, Brit. Foss. Corals, second series, part I (1866), pl. x, f. 20—22, p. 64. *Bracklesham Beds; Bracklesham.* Fisher Collection.

Beaumontia laxa (McCoy). *Columnaria laxa,* F. McCoy, A. M. N. H. ser. 2, III (1849), p. 122, and Contrib. Brit. Pal. (1854), p. 91, and Brit. Palæoz. Foss. (1851), pl. iii c, f. 11, p. 92. *Carboniferous Limestone; Derbyshire.* Presented by W. Hopkins, Esq.

Caninia lata, v. Omphyma turbinata.

C. subibicina, v. Zaphrentis? subibicina.

Cladochonus brevicollis, v. Monilopora (Cladochonus) brevicollis.

Clisiophyllum bipartitum, F. McCoy, A. M. N. H. ser. 2, III (1849), p. 2, and Contrib. Brit. Pal. (1854), p. 70, and Brit. Palæoz. Foss. (1851), pl. iii c, f. 6, 6 *a,* p. 93. H. M. Edwards and J. Haime, Archiv. Mus. Hist. Nat. v (1851), p. 411, and Brit. Foss. Corals (1852), p. 187. *Carboniferous Limestone; Derbyshire.* Presented by W. Hopkins, Esq.

C. coniseptum (Keyserling). *C. vortex,* F. McCoy, A. M. N. H. ser. 2, VI (1850), p. 277, and Contrib. Brit. Pal. (1854), p. 163, and Brit. Palæoz. Foss. (1851), pl. i B, f. 18, 18 *a,* p. 33.

Carboniferous Limestone ; locality unknown. Stated by McCoy to come from the Wenlock Limestone, of Wenlock.

Clisiophyllum ? costatum, v. Cyathaxonia costata.

C. keyserlingi, F. McCoy, A. M. N. H. ser. 2, III (1849), p. 2, and Contrib. Brit. Pal. (1854), p. 70, and Brit. Palæoz. Foss. (1851), pl. iii C, f. 4, 4 *a*, p. 94. *Carboniferous Limestone; Derbyshire.* Presented by W. Hopkins, Esq.

C. prolapsum, v. Cyclophyllum fungites.

C. turbinatum, F. McCoy, A. M. N. H. ser. 2, VII (1851), p. 169, and Contrib. Brit. Pal. (1854), p. 201, and Brit. Palæoz. Foss. (1851), p. 96, woodcuts *a, b, c*, on p. 88. *Carboniferous Limestone; Beith, Ayrshire.*

C. vortex, v. Clisiophyllum coniseptum.

Cœnites intertextus, Eichwald, H. M. Edwards and J. Haime, Archiv. Mus. Hist. Nat. v (1851), p. 302, and Brit. Foss. Corals (1854), pl. lxv, f. 5, 5 *a*, p. 276. *Wenlock Limestone; Dudley.* Fletcher Collection. There are several specimens in the collection, but none of them agrees well with the figure.

C. juniperinus, Eichwald, H. M. Edwards and J. Haime, Brit. Foss. Corals (1854), pl. lxv, f. 4, 4 *a*, p. 276. *Wenlock Limestone; Dudley.* Fletcher Collection.

C. labrosus, H. M. Edwards and J. Haime, Arch. Mus. Hist. Nat. v (1851), p. 302, and Brit. Foss. Corals (1854), pl. lxv, f. 6, 6 *a*, p. 277. *Wenlock Limestone; Dudley.* Fletcher Collection.

C. linearis, H. M. Edwards and J. Haime, Arch. Mus. Hist. Nat. v (1851), p. 302, and Brit. Foss. Corals (1854), pl. lxv, f. 3 p. 277. *Wenlock Limestone; Dudley.* Fletcher Collection.

C.? strigatus, F. McCoy, A. M. N. H. ser. 2, VI (1850), p. 476, and Contrib. Brit. Pal. (1854), p. 170, and Brit. Palæoz. Foss. (1851), pl. i C, f. 8, 8 *a*, p. 22. H. M. Edwards and J. Haime, Brit. Foss. Corals (1854), p. 278. *Wenlock Limestone; Dudley.*

C. sp. 1, Salter, Cat. (1873), p. 106. *Wenlock Limestone; Dudley.* Fletcher Collection.

C. sp. 2, Salter, Cat. (1873), p. 106. *Wenlock Limestone; Dudley.* Fletcher Collection.

Columnaria laxa, v. Beaumontia laxa.

Convexastræa waltoni, H. M. Edwards and J. Haime, Brit. Foss. Corals (1851), pl. xxiii, f. 5, 5 *a*, 6, p. 109. *Great Oolite; fig. 5 from Hampton Cliffs, fig. 6 from Box Tunnel, near Bath.* Walton Collection.

Cosmoseris vermicularis (McCoy). *Meandrina vermicularis,* F. McCoy, A. M. N. H. ser. 2, II (1848), p. 402, and Contrib. Brit. Pal. (1854), p. 49. *Inferior Oolite; Leckhampton.*

Cyathaxonia costata, F. McCoy, A. M. N. H. ser. 2, III (1849), p. 6, and Contrib. Brit. Pal. (1854), p. 74, and Brit. Palæoz. Foss. (1851), pl. iii C, f. 2, 2 *a*, p. 109. *Clisiophyllum?* costatum, H. M. Edwards and J. Haime, Arch. Mus. Hist. Nat. v (1851), p. 412, and Brit. Foss. Corals (1852), p. 187. *Carboniferous Limestone; Derbyshire.* Presented by W. Hopkins, Esq.

C. siluriensis, v. Lindströmia siluriensis.

Cyathophora luciensis (d'Orbigny), H. M. Edwards and J. Haime, Brit. Foss. Corals (1851), pl. xxx, f. 5, 5 *a*, p. 107. *Bradford Clay; Pound Hill.* Walton Collection. The specimens in the collection do not agree well with the figure.

Cyathophyllum angustum, Lonsdale. *Cystiphyllum brevilamellatum,* F. McCoy, A. M. N. H. ser. 2, VI (1850), p. 276, and Contrib. Brit. Pal. (1854), p. 162, and Brit. Palæoz. Foss. (1851), pl. i B, f. 19, 19 *a*, p. 32. *Wenlock Limestone; Wenlock.*

C. articulatum (Wahlenberg). H. M. Edwards and J. Haime, Brit. Foss. Corals (1854), pl. lxvii, f. 1, 1 *a*, p. 282. *Wenlock Limestone; Dudley.* Fletcher Collection.

C. craigense, v. Streptelasma craigense.

C. dianthoides, F. McCoy, A. M. N. H. ser. 2, III (1849), p. 7, and Contrib. Brit. Pal. (1854), p. 75, and Brit. Palæoz. Foss. (1851), pl. iii C, f. 7, 7 *a—c*, p. 85. H. M. Edwards and J. Haime, Arch. Mus. Hist. Nat. (1851), p. 390, and Brit. Foss. Corals (1852), p. 182. *Carboniferous Limestone; Arnside.*

C. murchisoni (Edwards and Haime). *Strephodes multilamellatum,* F. McCoy, A. M. N. H. ser. 2, III (1849), p. 5, and Contrib. Brit. Pal. (1854), p. 73, and Brit. Palæoz. Foss. (1851), pl. iii C, f. 3, 3 *a*, p. 93. *Carboniferous Limestone; Arnside, Westmorland.*

Cyathophyllum paracida, F. McCoy, A. M. N. H. ser. 2, III (1849), p. 7, and Contrib. Brit. Pal. (1854), p. 75, and Brit. Palæoz. Foss. (1851), pl. iii C, f. 9, p. 86. *Carboniferous Limestone; Derbyshire.*

C. pseudoceratites (McCoy), H. M. Edwards and J. Haime, Brit. Foss. Corals (1854), pl. lxvi, f. 3, 3 *a*, p. 282. *Wenlock Limestone; Dudley.* Fletcher Collection.
 Strephodes pseudoceratites, F. McCoy, A. M. N. H. ser. 2, VI (1850), p. 474, and Contrib. Brit. Pal. (1854), p. 159, and Brit. Palæoz. Foss. (1851), pl. i B, f. 20, 20 *a*, p. 30. *Woolhope Limestone; Old Radnor, Presteign.*

C.? pseudovermiculare, F. McCoy, A. M. N. H. ser. 2, III (1849), p. 8, and Contrib. Brit. Pal. (1854), p. 76, and Brit. Palæoz. Foss. (1851), pl. iii C, f. 8, 8 *a, b*, p. 86. H. M. Edwards and J. Haime, Arch. Mus. Hist. Nat. V (1851), p. 388, Brit. Foss. Corals (1852), p. 182. *Carboniferous Limestone; Kendal.*

C. regium, Phillips. *Astræa (Palastrœa) carbonaria*, F. McCoy, A. M. N. H. ser. 2, III (1849), p. 125, and Contrib. Brit. Pal. (1854), p. 94, and Brit. Palæoz. Foss. (1851), pl. iii A, f. 7, 7 *a*, pl. iii B, f. 1, p. 111. *Carboniferous Limestone; Derbyshire.* Presented by W. Hopkins, Esq.

C. trochiforme (McCoy), H. M. Edwards and J. Haime, Brit. Foss. Corals (1854), p. 285. *Strephodes trochiformis*, F. McCoy, A. M. N. H. ser. 2, VI (1850), p. 475, and Contrib. Brit. Pal. (1854), p. 160, and Brit. Palæoz. Foss. (1851), pl. i B, f. 21, p. 31. *Wenlock Limestone; Dudley.*

C. truncatum (Linnæus), H. M. Edwards and J. Haime, Brit. Foss. Corals (1854), pl. lxvi, f. 5, p. 284. *Wenlock Limestone; Dudley.* Fletcher Collection.
 Strephodes vermiculoides, F. McCoy, A. M. N. H. ser. 2, VI (1850), p. 275, and Contrib. Brit. Pal. (1854), p. 161, and Brit. Palæoz. Foss. (1851), pl. i B, f. 22, 22 *a*, p. 31. *Wenlock Limestone; near Aymestry.*

Cyathopsis? eruca, v. Lophophyllum? eruca.

Cyclophyllum fungites (Fleming). *Clisiophyllum prolapsum*, F. McCoy, A. M. N. H. ser. 2, III (1849), p. 3, and Contrib. Brit. Pal. (1854), p. 71, and Brit. Palæoz. Foss. (1851), pl. iii C,

f. 5, 5 *a*, p. 95. *Carboniferous Limestone; Derbyshire.* Presented by W. Hopkins, Esq.

Cystiphyllum brevilamellatum, v. Cyathophyllum angustum.

Cystiphyllum cylindricum, Lonsdale, H. M. Edwards and J. Haime, Brit. Foss. Corals (1854), pl. lxxii, f. 2 *b, c,* p. 297. *Wenlock Limestone; Dudley.* Fletcher Collection.

C. grayi, v. Actinocystis grayi.

C. sp., Salter, Cat. (1873), p. 116. *Wenlock Limestone; Dudley.* Fletcher Collection.

Dendrophyllia plicata, v. Goniocora socialis.

Dendropora megastoma, v. Rhabdopora megastoma.

Dentipora glomerata, v. Stylina tubulifera.

Diphyphyllum concinnum, Lonsdale. *D. lateseptatum,* F. McCoy, A. M. N. H. ser. 2, iii (1849), p. 8, and Contrib. Brit. Pal. (1854), p. 76, and Brit. Palæoz. Foss. (1851), pl. iii c, f. 10, 10 *a, b,* p. 88. *Carboniferous Limestone; Corwen.*

D. gracile, v. Lithostrotion irregulare.

D. lateseptatum, v. Diphyphyllum concinnum.

Discocyathus eudesi (Michelin), H. M. Edwards and J. Haime, Brit. Foss. Corals (1851), pl. xxv, f. 1, 1 *a, b,* p. 125. *Inferior Oolite; Burton Bradstock.* Walton Collection.

Favosites bowerbanki (Edwards and Haime). *Monticulipora? bowerbanki,* H. M. Edwards and J. Haime, Brit. Foss. Corals (1854), pl. lxiii, f. 1, 1 *a—c,* p. 268. *Wenlock Limestone; Dudley.* Fletcher Collection.

F. crassa, F. McCoy, A. M. N. H. ser. 2, vi (1850), p. 284, and Contrib. Brit. Pal. (1854), p. 170, and Brit. Palæoz. Foss. (1851), pl. i c, f. 9, 9 *a,* p. 20. H. M. Edwards and J. Haime, Brit. Foss. Corals (1854), p. 261. *Coniston Limestone; Coniston Water Head.*

F. forbesi, H. M. Edwards and J. Haime, Brit. Foss. Corals (1854), pl. lx, f. 2, 2 *a, b,* p. 258. *Wenlock Limestone; Dudley.* Fletcher Collection.

Fistulipora decipiens, F. McCoy, A. M. N. H. ser. 2, vi (1850), p. 285, and Contrib. Brit. Pal. (1854), p. 172, and Brit. Palæoz.

Foss. (1851), pl. i c, f. 1, 1 *a*, *b*, p. 11. H. M. Edwards and
J. Haime, Brit. Foss. Corals (1854), p. 298. *Wenlock Limestone; near Aymestry.*

Fistulipora dobunica, Nicholson and Foord. *Monticulipora sp.*
1 (*pars*) and *M. sp.* 2, Salter, Cat. (1873), pp. 108—9. *Wenlock
Limestone; Dudley.* Fletcher Collection.

F. incrustans (Phillips). *Fistulipora minor*, F. McCoy, A. M. N.
H. ser. 2, III (1849), p. 130, and Contrib. Brit. Pal. (1854), p. 99,
and Brit. Palæoz. Foss. (1851), pl. iii B, f. 12, 12 *a*, *b*, p. 79.
H. M. Edwards and J. Haime, Arch. Mus. Hist. Nat. V (1851),
p. 220, and Brit. Foss. Corals (1852), p. 151. *Carboniferous
Limestone; Derbyshire.* Presented by W. Hopkins, Esq.

F. major, F. McCoy, A. M. N. H. ser. 2, III (1849), p. 131, and
Contrib. Brit. Pal. (1854), p. 100. H. M. Edwards and J.
Haime, Arch. Mus. Hist. Nat. V (1851), p. 220, and Brit. Foss.
Corals (1852), p. 152. *Carboniferous Limestone; Derbyshire.*

F. minor, v. Fistulipora incrustans.

F. nummulina, Nicholson and Foord. *Monticulipora sp.* 1 (*pars*)
and *M. sp.* 3, Salter, Cat. (1873), pp. 108—9. *Wenlock Limestone; Dudley.* Fletcher Collection.

Goniocora socialis (Roemer). *Dendrophyllia plicata*, A. M. N.
H. ser. 2, II (1848), p. 403, and Contrib. Brit. Pal. (1854), p. 51.
Coral Rag; Steeple Ashton.

Goniophyllum fletcheri, H. M. Edwards and J. Haime, Arch.
Mus. Hist. Nat. V (1851), p. 405, and Brit. Foss. Corals (1854),
pl. lxviii, f. 3, 3 *a*, p. 290. *Wenlock Limestone; Dudley.*
Fletcher Collection.

Gonioseris angulata, P. M. Duncan, Brit. Foss. Corals, second
series, part iii (1872), pl. vii, f. 1—5, p. 21. *Millepore Bed;
Cloughton Wyke, near Scarborough.* Leckenby Collection.

G. leckenbyi, P. M. Duncan, Brit. Foss. Corals, second series,
part iii (1872), pl. vii, f. 6—9, p. 22. *Millepore Bed; Cloughton
Wyke, near Scarborough.* Leckenby Collection.

G., young form, P. M. Duncan, Brit. Foss. Corals, second series,
part iii (1872), pl. vii, f. 10, 11, p. 21. *Millepore Bed;
Cloughton Wyke, near Scarborough.* Leckenby Collection.

Heliolites cæspitosa, Salter, Cat. (1873), p. 104. *Wenlock Limestone; Dudley.* Fletcher Collection.

H. megastoma (McCoy). *Palæopora megastoma*, F. McCoy, Brit. Palæoz. Foss. (1851), pl. i C, f. 4, 4 *a, b*, p. 16. *Coniston Limestone; Coniston.*

H. murchisoni, Edwards and Haime. *Palæopora interstincta* (Linnæus) var. *subtubulata*, F. McCoy, Brit. Palæoz. Foss. (1851), pl. i C, f. 2, 2 *a, b*, p. 16. *Coniston Limestone; Coniston Water Head.*

Heterophyllia grandis, F. McCoy, A. M. N. H. ser. 2, III (1849), p. 126, and Contrib. Brit. Pal. (1854), p. 95, and Brit. Palæoz. Foss. (1851), pl. iii A, f. 1, 1 *a*, p. 112. H. M. Edwards and J. Haime, Arch. Mus. Hist. Nat. V (1851), p. 467, and Brit. Foss. Corals (1852), p. 210. *Carboniferous Limestone; Derbyshire.* Presented by W. Hopkins, Esq.

H. ornata, F. McCoy, A. M. N. H. ser. 2, III (1849), p. 127, and Contrib. Brit. Pal. (1854), p. 96, and Brit. Palæoz. Foss. (1851), pl. iii A, f. 2, 2 *a*, p. 112. H. M. Edwards and J. Haime, Arch. Mus. Nat. Hist. V (1851), p. 467, and Brit. Foss. Corals (1852), p. 210. *Carboniferous Limestone; Derbyshire.* Presented by W. Hopkins, Esq.

Isastræa explanulata (McCoy). *Astræa explanulata*, F. McCoy, A. M. N. H. ser. 2, II (1848), p. 400, and Contrib. Brit. Pal. (1854), p. 48. *Inferior Oolite; Dundry and Bath.*

I. stricklandi, P. M. Duncan, Brit. Foss. Corals, second series, part iv (1867), pl. xiii, f. 1—4, p. 54. *Lower Lias; Chadbury, Worcestershire.* Strickland Collection.

I. tenuistriata (McCoy). *Astræa tenuistriata*, F. McCoy, A. M. N. H. ser. 2, II (1848), p. 400, and Contrib. Brit. Pal. (1854), p. 47. *Inferior Oolite; Dundry.*

Lepidophyllia stricklandi, P. M. Duncan, Brit. Foss. Corals, second series, part iv (1867), pl. xii, f. 15, p. 53. *Lower Lias; Chadbury, Worcestershire.* Strickland Collection.

Lindströmia siluriensis (McCoy). *Cyathaxonia? siluriensis*, F. McCoy, A. M. N. H. ser. 2, V (1850), p. 281, and Contrib. Brit. Pal. (1854), p. 166, and Brit. Palæoz. Foss. (1851), pl. i C,

f. 11, 11 *a*, p. 36. H. M. Edwards and J. Haime, Brit. Foss. Corals (1854), p. 279. *Lower? Ludlow; Underbarrow.*

Lindströmia subduplicata (McCoy). *Petraia subduplicata,* F. McCoy, A. M. N. H. ser. 2, VI (1850), p. 279, and Contrib. Brit. Pal. (1854), p. 165, and Brit. Palæoz. Foss. (1851), pl. i B, f. 26, 26 *a*, p. 40. *May Hill Group; Dalquharran, Ayrshire.*

L. subduplicata var. **crenulata** [*Petraia,* McCoy], F. McCoy, A. M. N. H. ser. 2, VI (1850), p. 280, and Contrib. Brit. Pal. (1854), p. 165, and Brit. Palæoz. Foss. (1851), pl. i B, f. 26 *b*, p. 41. *May Hill Group; Dalquharran.*

L. uniserialis (McCoy). *Petraia uniserialis,* F. McCoy, A. M. N. H. ser. 2, VI (1850), p. 280, and Contrib. Brit. Pal. (1854), p. 165, and Brit. Palæoz. Foss. (1851), pl. i B, f. 25, 25 *a*, p. 41. *Upper Bala; Llansaintffraid.*

L. uniserialis var. **gràcilis** [*Petraia,* McCoy], F. McCoy, A. M. N. H. ser. 2, VI (1850), p. 281, and Contrib. Brit. Pal. (1854), p. 166, and Brit. Palæoz. Foss. (1851), p. 41. *Upper Bala; Llansaintffraid, etc.*

Lithostrotion arachnoideum (McCoy), H. M. Edwards and J. Haime, Brit. Foss. Corals (1852), p. 202. *Nemaphyllum arachnoideum,* F. McCoy, A. M. N. H. ser. 2, III (1849), p. 16, and Contrib. Brit. Pal. (1854), p. 84. and (*Nematophyllum*) Brit. Palæoz. Foss. (1851), pl. iii A, f. 6, 6 *a, b,* p. 97. *Stylaxis arachnoidea,* H. M. Edwards and J. Haime, Arch. Mus. Hist. Nat. V (1851), p. 454. *Carboniferous Limestone; Derbyshire.*

L. basaltiforme (Conybeare and Phillips). *Nemaphyllum minus,* F. McCoy, A. M. N. H. ser. 2, III (1849), p. 17, and Contrib. Brit. Pal. (1854), p. 85, and (*Nematophyllum*) Brit. Palæoz. Foss. (1851), pl. iii B, f. 3, 3 *a, b,* p. 99. *Carboniferous Limestone; Kendal.*

L. decipiens (McCoy), H. M. Edwards and J. Haime, Arch. Mus. Hist. Nat. V (1851), p. 441, and Brit. Foss. Corals (1852), p. 196. *Nemaphyllum decipiens,* F. McCoy, A. M. N. H. ser. 2, III (1849), p. 18, and Contrib. Brit. Pal. (1854), p. 86, and (*Nematophyllum*) Brit. Palæoz. Foss. (1851), p. 99. *Carboniferous Limestone; Derbyshire.*

L.? derbiense, Edwards and Haime. *Stylastrœa irregularis,*

F. McCoy, A. M. N. H. ser. 2, III (1849), p. 9, and Contrib.
Brit. Pal. (1854), p. 77. *Stylaxis irregularis*, F. McCoy, Brit.
Palæoz. Foss. (1851), pl. iii A, f. 5, 5 *a*, *b*, p. 101. *Carboniferous
Limestone; Derbyshire.* Presented by W. Hopkins, Esq.

Lithostrotion flemingi (McCoy), H. M. Edwards and J. Haime,
Brit. Foss. Corals (1852), p. 203. *Stylaxis flemingi*, F. McCoy,
A. M. N. H. ser. 2, III (1849), p. 121, and Contrib. Brit. Pal.
(1854), p. 89, and Brit. Palæoz. Foss. (1851), pl. iii A, f. 3, 3 *a*,
b, p. 100. H. M. Edwards and J. Haime, Arch. Mus. Hist.
Nat. v (1851), p. 454. *Carboniferous Limestone; Derbyshire.*
Presented by W. Hopkins, Esq.

L. irregulare (Phillips), H. M. Edwards and J. Haime, Brit.
Foss. Corals (1852), p. 199. *Diphyphyllum gracile*, F. McCoy,
A. M. N. H. ser. 2, VII (1851), p. 168, and Contrib. Brit. Pal.
(1854), p. 200, and Brit. Palæoz. Foss. (1851), p. 88, woodcuts *d*,
e, *f*. *Carboniferous Limestone; Lowick.*

L. major (McCoy), H. M. Edwards and J. Haime, Brit. Foss.
Corals (1852), p. 201. *Stylaxis major*, F. McCoy, A. M. N. H.
ser. 2, III (1849), p. 120, and Contrib. Brit. Pal. (1854), p. 89,
and Brit. Palæoz. Foss. (1851), pl. iii A, f. 4, 4 *a*, *b*, p. 101. H.
M. Edwards and J. Haime, Arch. Mus. Hist. Nat. v (1851),
p. 454. *Carboniferous Limestone; Derbyshire.* Presented by
W. Hopkins, Esq.

L. portlocki (Bronn). *Nemaphyllum clisioides*, A. M. N. H.
ser. 2, III (1849), p. 18, and Contrib. Brit. Pal. (1854), p. 86,
and (*Nematophyllum*) Brit. Palæoz. Foss. (1851), pl. iii B,
f. 2, 2 *a*, *b*, p. 98. *Carboniferous Limestone; Derbyshire.*
Presented by W. Hopkins, Esq.

L.? septosum (McCoy), H. M. Edwards and J. Haime, Arch.
Mus. Hist. Nat. v (1851), p. 444, and Brit. Foss. Corals (1852),
p. 196. *Nemaphyllum septosum*, F. McCoy, A. M. N. H. ser. 2,
III (1849), p. 19, and Contrib. Brit. Pal. (1854), p. 87. *Carboniferous Limestone; Tullyard, Armagh.*

Lonsdaleia crassiconus, v. Lonsdaleia duplicata.

Lonsdaleia duplicata (Martin), H. M. Edwards and J. Haime,
Brit. Foss. Corals (1852), p. 209. *Lonsdaleia stylastræformis*,
F. McCoy, A. M. N. H. ser. 2, III (1849), p. 14, and Contrib.

Brit. Pal. (1854), p. 82, and Brit. Palæoz. Foss. (1851), pl. iii B, f. 7, 7 *a*, *b*, p. 106. *Carboniferous Limestone; Kendal.*

Lonsdaleia crassiconus, F. McCoy, A. M. N. H. ser. 2, III (1849), p. 12, and Contrib. Brit. Pal. (1854), p. 80, and Brit. Palæoz. Foss. (1851), pl. iii B, f. 5, 5 *a*, *b*, p. 104. H. M. Edwards and J. Haime, Arch. Mus. Hist. Nat. V (1851), p. 461. *Carboniferous Limestone; Arnside.*

Lonsdaleia floriformis (Martin). *Strombodes conaxis*, F. McCoy, A. M. N. H. ser. 2, III (1849), p. 10, and Contrib. Brit. Pal. (1854), p. 78, and Brit. Palæoz. Foss. (1851), pl. iii B, f. 4, 4 *a*, *b*, p. 102. *Carboniferous Limestone; Derbyshire.* Presented by W. Hopkins, Esq.

L. rugosa, F. McCoy, A. M. N. H. ser. 2, III (1849), p. 13, and Contrib. Brit. Pal. (1854), p. 81, and Brit. Palæoz. Foss. (1855), pl. iii B, f. 6, 6 *a*, *b*, p. 105. *Carboniferous Limestone; Corwen.*

L. stylastræformis, v. Lonsdaleia duplicata.

L. wenlockensis (McCoy), H. M. Edwards and J. Haime, Brit. Foss. Corals (1854), p. 296. *Strombodes wenlockensis*, F. McCoy, A. M. N. H. ser. 2, VI (1850), p. 274, and Contrib. Brit. Pal. (1854), p. 158, and Brit. Palæoz. Foss. (1851), pl. i B, f. 28, 28 *a*, p. 34. *Carboniferous Limestone; locality unknown.* Stated by McCoy to come from the Wenlock Limestone, near Wenlock, Shropshire.

Lophophyllum? eruca (McCoy), H. M. Edwards and J. Haime, Brit. Foss. Corals (1852), p. 177. *Cyathopsis? eruca*, F. McCoy, A. M. N. H. ser. 2, VII (1851), p. 167, and Contrib. Brit. Pal. (1854), p. 199, and Brit. Palæoz. Foss. (1851), pl. iii I, f. 34, 34 *a*, *b*, p. 90. *Carboniferous Limestone; Beith, Ayrshire.*

Lyopora favosa (McCoy), H. A. Nicholson and R. Etheridge, jun., Mon. Sil. Foss. Girvan, I, fascic. I (1878), p. 27. *Palæopora favosa*, F. McCoy, A. M. N. H. ser. 2, VI (1850), p. 285, and Contrib. Brit. Pal. (1854), p. 171, and Brit. Palæoz. Foss. (1851), pl. i C, f. 3, 3 *a*, *b*, p. 15. *Heliolites interstincta* (Lin.), H. M. Edwards and J. Haime, Brit. Foss. Corals (1854), p. 250. *Middle Bala; Craig Head, Girvan.*

Meandrina vermicularis, v. Cosmoseris vermicularis.

Michelinia glomerata, v. Michelinia tenuisepta.

M. grandis, v. Michelinia megastoma.

Michelinia megastoma (Phillips). *M. grandis,* F. McCoy, A. M.
N. H. ser. 2, III (1849), p. 123, and Contrib. Brit. Pal. (1854),
p. 92, and Brit. Palæoz. Foss. (1851), pl. iii C, f. 1, 1 *a*, p. 81.
Carboniferous Limestone; Arnside.

M. tenuisepta (Phillips). *M. glomerata,* F. McCoy, A. M. N. H.
ser. 2, III (1849), p. 122, and Contrib. Brit. Pal. (1854), p. 91,
and Brit. Palæoz. Foss. (1851), pl. iii B, f. 14, p. 80. H. M.
Edwards and J. Haime, Brit. Foss. Corals (1852), p. 155.
Carboniferous Limestone; Derbyshire. Presented by W. Hopkins, Esq.

Microsolena excelsa, H. M. Edwards and J. Haime, Brit. Foss.
Corals (1851), pl. xxv, f. 5, 5 *a*, p. 124. *Great Oolite; near
Bath.* Walton Collection.

M. regularis, H. M. Edwards and J. Haime, Brit. Foss. Corals
(1851), pl. xxv, f. 6, 6 *a, b,* p. 122. *Great Oolite; Bathford Hill.*
Walton Collection.

Monilopora (Cladochonus) brevicollis (McCoy), H. A. Nicholson and R. Etheridge, jun., G. M. dec. 2, VI (1879), p. 293.
H. A. Nicholson, Tabul. Corals (1879), p. 222. *Cladochonus
brevicollis,* F. McCoy, A. M. N. H. ser. 2, III (1849), p. 128,
and Contrib. Brit. Pal. (1854), p. 164, and Brit. Palæoz. Foss.
(1851), pl. iii B, f. 10, p. 85. *Syringopora sp.* [young], H. M.
Edwards and J. Haime, Arch. Mus. Hist. Nat., V (1851), p. 296,
and Brit. Foss. Corals (1852), p. 164. *Carboniferous Limestone;
Derbyshire.* Presented by W. Hopkins, Esq.

Monotrypella pulchella (Edwards and Haime). *Monticulipora
pulchella,* H. M. Edwards and J. Haime, Brit. Foss. Corals
(1854), pl. lxii, f. 5, 5 *a, b,* p. 267. *Wenlock Limestone; Dudley.*
Fletcher Collection.

M. sp. 1 (*pars*) and *M. sp.* 4, Salter, Cat. (1873), pp. 108—9.
Wenlock Limestone; Dudley. Fletcher Collection.

Monticulipora bowerbanki, v. Favosites bowerbanki.

Monticulipora? explanata (McCoy), H. M. Edwards and J.
Haime, Brit. Foss. Corals (1854), p. 268. *Nebulipora explanata,*

F. McCoy, A. M. N. H. ser. 2, VI (1850), p. 283, and Contrib.
Brit. Pal. (1854), p. 168, and Brit. Palæoz. Foss. (1851), pl. i c,
f. 6, 6 *a*, p. 23. *Coniston Limestone; Coniston.*

Monticulipora? lens (McCoy), H. M. Edwards and J. Haime,
Brit. Foss. Corals (1854), p. 269. *Nebulipora lens*, F. McCoy,
A. M. N. H. ser. 2, VI (1850), p. 283, and Contrib. Brit. Pal.
(1854), p. 169, and Brit. Palæoz. Foss. (1851), pl. i c, f. 7, 7 *a, b,*
p. 23. *Middle Bala; Horderley West.*

M.? papillata (McCoy). *Nebulipora papillata*, F. McCoy, A. M.
N. H. ser. 2, VI (1850), p. 284, and Contrib. Brit. Pal. (1854),
p. 169, and Brit. Palæoz. Foss. (1851), pl. i c, f. 5, 5 *a*, p. 24.
Coniston Limestone; Coniston.

M.? poculum, Salter, Cat. (1873), p. 109. *Wenlock Limestone;*
Dudley. Fletcher Collection.

M. pulchella, v. Monotrypella pulchella.

M. sp. 1 (Salter), v. Fistulipora dobunica, F. nummulina, and
Monotrypella pulchella.

M. sp. 2 (Salter), v. Fistulipora dobunica.

M. sp. 3 (Salter), v. Fistulipora nummulina.

M. sp. 4 (Salter), v. Monotrypella pulchella.

M. sp. 5, Salter, Cat. (1873), p. 109. *Wenlock Limestone; Dudley.*
Fletcher Collection.

M. sp. 6, Salter, Cat. (1873), p. 109. *Wenlock Limestone; Dudley.*
Fletcher Collection.

Montlivaltia depressa, H. M. Edwards and J. Haime, Brit. Foss.
Corals (1851), pl. xxix, f. 5, 5 *a*, p. 134. *Inferior Oolite;*
Wootton-under-Edge. Walton Collection. This specimen cannot
be found.

M. gregaria, v. Thecosmilia gregaria.

M. lens, H. M. Edwards and J. Haime, Brit. Foss. Corals (1851),
pl. xxvi, f. 7, 7 *a, b*, 8, p. 133. *Inferior Oolite; Charlcombe.*
Walton Collection. These specimens cannot be found.

M. tenuilamellosa, H. M. Edwards and J. Haime, Brit. Foss.
Corals (1851), pl. xxvi, f. 11, 11 *a*, p. 130. *Fuller's Earth;*
Dunkerton, Somerset. Walton Collection.

Montlivaltia trochoides, H. M. Edwards and J. Haime, Brit. Foss. Corals (1851), pl. xxvi, f. 2, 2 *a*, 3, 3 *a*, 4, 10; pl. xxvii, f. 2, 2 *a*, p. 129. *Fuller's Earth ; Charlcombe.* Walton Collection. Fig. 3 of pl. xxvi, and fig. 2 of pl. xxvii, cannot be found.

Nebulipora explanata, v. Monticulipora? explanata.

N. lens, v. Monticulipora lens.

N. papillata, v. Monticulipora papillata.

Nemaphyllum septosum, v. Lithostrotion? septosum.

Nematophyllum arachnoideum, v. Lithostrotion arachnoideum.

N. clisioides, v. Lithostrotion portlocki.

N. decipiens, v. Lithostrotion decipiens.

N. minus, v. Lithostrotion basaltiforme.

Nidulites favus, Salter. *Palæopora favosa*, F. McCoy (*pars*), A. M. N. H. ser. 2, vi (1850), p. 285, and Contrib. Brit. Pal. (1854), p. 171, and Brit. Palæoz. Foss. (1851), pl. i c, f. 3 c, d, p. 15. *May Hill Group ; Mullock Quarry, Dalquharran.*

Omphyma turbinata (Fougt). *Caninia lata*, F. McCoy, A. M. N. H. ser. 2, vi (1850), p. 277, and Contrib. Brit. Pal. (1854), p. 162, and Brit. Palæoz. Foss. (1851), pl. i c, f. 12, 13, p. 28. *Wenlock Limestone ; Wenlock.*

Palæocyclus fletcheri, H. M. Edwards and J. Haime, Arch. Mus. Hist. Nat. v (1851), p. 205, and Brit. Foss. Corals (1854), pl. lxvii, f. 3, 3 *a—f*, p. 248. *Wenlock Limestone ; Dudley.* Fletcher Collection.

P. porpita (Linnæus), H. M. Edwards and J. Haime, Brit. Foss. Corals (1854), pl. lvii, f. 1, 1 *a—c*, p. 246. *Wenlock Limestone ; Dudley.* Fletcher Collection.

P. rugosus, H. M. Edwards and J. Haime, Brit. Foss. Corals (1854), pl. lvii, f. 4, 4 *a—d*, p. 248. *Wenlock Limestone ; Dudley.* Fletcher Collection.

Palæopora favosa, v. Lyopora favosa and Nidulites favus.

P. interstincta var. *subtubulata*, v. Heliolites murchisoni.

P. megastoma, v. Heliolites megastoma.

Palæopora subtilis, F. McCoy, A. M. N. H. ser. 2, vi (1850), p. 476, and Contrib. Brit. Pal. (1854), p. 172, and Brit. Palæoz. Foss.

ACTINOZOA. 27

(1851), p. 17. H. A. Nicholson and R. Etheridge, jun., Mon. Sil. Foss. Girvan, I, fascic. III (1880), p. 253. *May Hill Group; Dalquharran, Ayrshire.*

Petraia æquisulcata, v. Streptelasma æquisulcatum.

Petraia celtica (Lamouroux), W. Lonsdale, Trans. Geol. Soc. ser. 2, V (1840), pl. lxviii, f. 6. *Devonian; Padstow.*

P.? **gigas**, F. McCoy, A. M. N. H. ser. 2, III (1849), p. 1, and Contrib. Brit. Pal. (1854), p. 69, and Brit. Palæoz. Foss. (1851), p. 74, woodcuts *c—e* on p. 66. *Cyathophyllum bucklandi,* H. M. Edwards and J. Haime, Arch. Mus. Hist. Nat. V (1851), p. 390, and Brit. Foss. Corals (1853), p. 226. *Devonian; New Quay, Cornwall.*

P. subduplicata, v. Lindströmia subduplicata.

P. uniserialis, v. Lindströmia uniserialis.

Phillipsastræa radiata (S. Woodward), H. M. Edwards and J. Haime, Brit. Foss. Corals (1852), pl. xxxvii, f. 2, 2 *a*, p. 203. R. Schäfer, G. M. dec. 3, VI (1889), p. 401, woodcuts 1—3. *Sarcinula phillipsi,* F. McCoy, A. M. N. H. ser. 2, III (1849), p. 125, and Contrib. Brit. Pal. (1854), p. 94, and Brit. Palæoz. Foss. (1851), p. 110. *Sarcinula placenta,* F. McCoy, A. M. N. H. ser. 2, III (1849), p. 124, and Contrib. Brit. Pal. (1854), p. 93, and Brit. Palæoz. Foss. (1851), pl. iii B, f. 9, 9 *a, b,* p. 110. *Carboniferous Limestone; Corwen and Derbyshire.*

P. tuberosa (McCoy), H. M. Edwards and J. Haime, Arch. Mus. Hist. Nat. V (1851), p. 449, and Brit. Foss. Corals (1852), p. 204. R. Schäfer, G. M. dec. 3, VI (1889), p. 407, woodcuts 4—6. *Sarcinula tuberosa,* F. McCoy, A. M. N. H. ser. 2, III (1849), p. 124, and Contrib. Brit. Pal. (1854), p. 93, and Brit. Palæoz. Foss. (1851), pl. iii B, f. 8, 8 *a,* p. 110. *Carboniferous Limestone; Derbyshire.*

Protoseris waltoni, H. M. Edwards and J. Haime, Brit. Foss. Corals (1851), pl. xx, p. 103. *Coral Rag; Osmington.* Walton Collection.

Rhabdopora megastoma (McCoy), H. M. Edwards and J. Haime, Arch. Mus. Hist. Nat. V (1851), p. 405, and Brit. Foss. Corals (1852), p. 165. *Dendropora megastoma,* F. McCoy,

A. M. N. H. ser. 2, III (1849), p. 129, and Contrib. Brit. Pal.
(1854), p. 98, and Brit. Palæoz. Foss. (1851), pl. iii B, f. 11, 11 a,
p. 79.   *Carboniferous Limestone; Derbyshire.*

Sarcinula phillipsi, v. Phillipsastræa radiata.

S. placenta, v. Phillipsastræa radiata.

S. tuberosa, v. Phillipsastræa tuberosa.

Siderastrœa agariciformis, v. Thamnastræa arachnoides.

Strephodes craigensis, v. Streptelasma craigense.

Strephodes gracilis, F. McCoy, A. M. N. H. ser. 2, VI (1850),
p. 378, and Brit. Palæoz. Foss. (1851), p. 72.   *Cyathophyllum*
or *Ptychophyllum,* H. M. Edwards and J. Haime, Brit. Foss.
Corals (1854), p. 232.   *Devonian; Newton Bushel.*

S. multilamellatum, v. Cyathophyllum murchisoni.

S. pseudoceratites, v. Cyathophyllum pseudoceratites.

S. trochiformis, v. Cyathophyllum trochiforme.

S. vermiculoides, v. Cyathophyllum truncatum.

Streptelasma æquisulcatum (McCoy), H. A. Nicholson and
R. Etheridge, jun., Mon. Sil. Foss. Girvan, I, fascic. I (1878),
p. 80.   *Petraia æquisulcata,* F. McCoy, A. M. N. H. ser. 2, VI
(1850), p. 279, and Contrib. Brit. Pal. (1854), p. 164, and
Brit. Palæoz. Foss. (1851), pl. i B, f. 23, 24, p. 39.   *Coniston
Limestone; Coniston.*

S. craigense (McCoy).   *Strephodes craigensis,* F. McCoy, A. M.
N. H. ser. 2, VI (1850), p. 275, and Contrib. Brit. Pal. (1854),
p. 129, and Brit. Palæoz. Foss. (1851), pl. i C, f. 10, 10 a, b,
p. 30.   *Cyathophyllum craigense,* Salter, Cat. (1873), p. 43.
Middle Bala; Craig Head, Girvan.

Strombodes conaxis, v. Lonsdaleia floriformis.

S. wenlockensis, v. Lonsdaleia wenlockensis.

Stylastrœa irregularis, v. Lithostrotion? derbiense.

Stylaxis flemingi, v. Lithostrotion flemingi.

S. irregularis, v. Lithostrotion? derbiense.

S. major, v. Lithostrotion major.

Stylina ploti, H. M. Edwards and J. Haime, Brit. Foss. Corals (1851), pl. xxiii, f. 1, p. 106. *Great Oolite; Combe Down, near Bath.* Walton Collection.

S. solida (McCoy). *Stylopora solida,* F. McCoy, A. M. N. H. ser. 2, II (1848), p. 399, and Contrib. Brit. Pal. (1854), p. 47. *Inferior Oolite; Dundry.*

S. tubulifera (Phillips). *Dentipora glomerata,* F. McCoy, A. M. N. H. ser. 2, II (1848), p. 399, and Contrib. Brit. Pal. (1854), p. 47. *Coral Rag; Malton.*

Stylopora solida, v. Stylina solida.

Syringopora bifurcata, Lonsdale, H. M. Edwards and J. Haime, Brit. Foss. Corals (1854), pl. lxiv, f. 3, 3 *a, b,* p. 273. *Wenlock Limestone; Dudley.* Fletcher Collection.

S. fascicularis (Linnæus), H. M. Edwards and J. Haime, Brit. Foss. Corals (1854), pl. lxv, f. 1 *b,* p. 274. *Wenlock Limestone; Dudley.* Fletcher Collection.

S. serpens (Linnæus), H. M. Edwards and J. Haime, Brit. Foss. Corals (1854), pl. lxv, f. 2, 2 *a,* p. 275. *Wenlock Limestone; Dudley.* Fletcher Collection.

Thamnastrea arachnoides (Parkinson). *Siderastrœa agariciformis,* F. McCoy, A. M. N. H. ser 2, II (1848), p. 401, and Contrib. Brit. Pal. (1854), p. 49. *Coral Rag; Upware.*

T. defranciana (Michelin). *Astrœa helianthella,* A. M. N. H. ser. 2, II (1848), p. 401, and Contrib. Brit. Pal. (1854), p. 48. H. M. Edwards and J. Haime, Brit. Foss. Corals (1851), p. 140. *Inferior Oolite; Dundry.*

T. fungiformis, H. M. Edwards and J. Haime, Brit. Foss. Corals (1851), pl. xxx, f. 4, 4 *a,* p. 141. *Inferior Oolite; Charlcombe.* Walton Collection.

T. scita, H. M. Edwards and J. Haime, Brit. Foss. Corals (1851), pl. xxiii, f. 4, 4 *a,* p. 119. *Great Oolite; Hampton Cliffs, near Bath.* Walton Collection.

T. waltoni, H. M. Edwards and J. Haime, Brit. Foss. Corals (1851), pl. xxv, f. 4, 4 *a,* p. 120. *Great Oolite; Box Tunnel, near Bath.* Walton Collection.

Thecosmilia annularis (Fleming). '*A Madreporite*' J. Parkinson, Organic Remains, II (1808), pl. v, f. 5, p. 67.   *Coral Rag;* *Steeple Ashton.*

T. gregaria (McCoy). *Montlivaltia gregaria*, F. McCoy, A. M. N. H. ser. 2, II (1848), p. 403, and Contrib. Brit. Pal. (1854), p. 51.   *Inferior Oolite; Cheltenham and Dundry.*

Trochocyathus conulus (Michelin), A. J. Jukes-Browne, Q. J. G. S. XXXI (1875), pl. xiv, f. 14—16, p. 303.   *Cambridge Greensand; Cambridge.*

Zaphrentis? subibicina (McCoy), H. M. Edwards and J. Haime, Brit. Foss. Corals (1852), p. 172.   *Caninia subibicina*, F. McCoy, A. M. N. H. ser. 2, VII (1851), p. 167, and Contrib. Brit. Pal. (1854), p. 199, and Brit. Palæoz. Foss. (1851), pl. iii I, f. 35, 35 *a*, p. 89.   *Carboniferous Limestone; Kendal.*

Z.? waltoni, H. M. Edwards and J. Haime, Brit. Foss. Corals (1851), pl. xxvii, f. 8, 8 *a*, p. 143.   P. M. Duncan, Rep. Brit. Assoc. (1869), p. 161.   *Inferior Oolite; Dundry.*   Probably derived from the Carboniferous Limestone.   Walton Collection.

ECHINODERMATA.

ECHINOIDEA.

Acrosalenia rarispina, v. Hemicidaris minor.

Acrosalenia sp. *Hemicidaris confluens*, F. McCoy, A. M. N. H. ser. 2, II (1848), p. 411.   T. Wright, Brit. Foss. Echin. Oolit. (1857), p. 92.   *Great Oolite; Minchinhampton.*

Arbacia inflata, v. Cottaldia sp.

Archæocidaris urei (Fleming), W. Keeping, Q. J. G. S. XXXII (1876), pl. iii, f. 14—18, p. 39.   *Carboniferous Limestone; Hafod-y-calch, Corwen.*

Cardiaster suborbicularis, v. Holaster suborbicularis.

Cidaris gaultina, Forbes, A. J. Jukes-Browne, Q. J. G. S., XXXI (1875), pl. xv, f. 13, p. 302.   *Cambridge Greensand; Cambridge.*

Cidaris sedgwicki, H. G. Seeley, A. M. N. H. ser. 3, VIII (1861), p. 22. *Cambridge Greensand; Cambridge.*

C. sp., W. Keeping, Foss. Upware and Brickhill (1883), pl. vii, f. 9, pp. 133—4. *Lower Greensand; Upware.*

Clypeus altus, F. McCoy, A. M. N. H. ser. 2, II (1848), p. 417, and Contrib. Brit. Pal. (1854), p. 65. *Inferior Oolite; Bridport.*

C. excentricus, v. Clypeus sinuatus.

C. sinuatus, Leske. *C. excentricus*, F. McCoy, A. M. N. H. ser. 2, II (1848), p. 417, and Contrib. Brit. Pal. (1854), p. 64. *Inferior Oolite; Leckhampton.*

Collyrites ovalis (Leske). *Dysaster symmetricus*, F. McCoy, A. M. N. H. ser. 2, II (1848), p. 414, and Contrib. Brit. Pal. (1854), p. 62. *Inferior Oolite; Bridport.*

Cottaldia sp. *Arbacia inflata*, F. McCoy, A. M. N. H. ser. 2, II (1848), p. 410, and Contrib. Brit. Pal. (1854), p. 58. *Cambridge Greensand; Cambridge.* This specimen is missing.

Cyphosoma? impressa, v. Echinocyphus impressus.

Diadema fungoideum, v. Pseudodiadema fungoideum.

D. intertuberculatum, v. Pseudodiadema intertuberculatum.

D. (Tetragamma?) scriptum, v. Diplopodia scriptum.

Diplopodia pentagona, F. McCoy, A. M. N. H. ser. 2, II (1848), p. 412, and Contrib. Brit. Pal. (1854), p. 60. T. Wright, Brit. Foss. Echin. Oolit. (1855), pl. vi, f. 3 *a, b, c, d*, p. 115. *Great Oolite; Minchinhampton.*

D. scriptum (Seeley). *Diadema (Tetragramma?) scriptum*, H. G. Seeley, A. M. N. H. ser. 3, VIII (1861), p. 20. *Cambridge Greensand; Cambridge.*

Discoidea marginalis, v. Holectypus hemisphæricus.

Discoidea minima, Desor, F. McCoy, A. M. N. H. ser. 2, II (1848), p. 420, and Contrib. Brit. Pal. (1854), p. 67. T. Wright, Brit. Foss. Echin. Cret. I (1874), p. 210. *Chalk; locality doubtful.*

Dysaster symmetricus, v. Collyrites ovalis.

Echinocyphus impressus (Seeley). *Cyphosoma? impressa*, H. G. Seeley, A. M. N. H. ser. 3, VIII (1861), p. 18. *Cambridge Greensand; Cambridge.*

Echinus diademata, v. Stomechinus germinans.

E. petallatus, v. Stomechinus gyratus.

Galeropygus agariciformis (Forbes). *Pygaster sublœvis*, F. McCoy, A. M. N. H. ser. 2, II (1848), p. 413, and Contrib. Brit. Pal. (1854), p. 61. *Inferior Oolite; Leckhampton.*

Goniophorus lunatus, Agassiz, var. **minutus**, H. G. Seeley, A. M. N. H. ser. 3, VIII (1861), p. 17. *Cambridge Greensand; Cambridge.* Presented by J. Carter, Esq., F.G.S.

Hemiaster m'coyi, H. G. Seeley, A. M. N. H. ser. 3, VIII (1861), p. 16. *Cambridge Greensand; Cambridge.*

Hemicidaris confluens, v. Acrosalenia sp.

Hemicidaris minor, Agassiz. *Acrosalenia rarispina*, F. McCoy, A. M. N. H. ser. 2, II (1848), p. 411, and Contrib. Brit. Pal. (1854), p. 59. *Great Oolite; Minchinhampton?*

Holaster suborbicularis (Brongniart). *Cardiaster suborbicularis*, H. G. Seeley, A. M. N. H. ser. 3, XVII (1866), p. 179. *Red Chalk; Hunstanton.*

Holectypus hemisphæricus (Agassiz). *Discoidea marginalis*, F. McCoy, A. M. N. H. ser. 2, II (1848), p. 413, and Contrib. Brit. Pal. (1854), p. 60. *Inferior Oolite; Broadwinsor, Dorset.*

Nucleolites æqualis, F. McCoy, A. M. N. H. ser. 2, II (1848), p. 416, and Contrib. Brit. Pal. (1854), p. 64. *Inferior Oolite; Castle Ashby.*

N. clunicularis (Phillips). *Nucleolites pyramidalis*, F. McCoy, A. M. N. H. ser. 2, II (1848), p. 416, and Contrib. Brit. Pal. (1854), p. 63. T. Wright, Brit. Foss. Echin. Oolit. (1859), p. 335. *Coral Rag; Weymouth.*

N. planulatus, F. McCoy, A. M. N. H. ser. 2, II (1848), p. 415, and Contrib. Brit. Pal. (1854), p. 63. *Great Oolite; Minchinhampton.*

N. pyramidalis, v. Nucleolites clunicularis.

Palæechinus gigas, McCoy, W. Keeping, Q. J. G. S. XXXII

(1876), pl. iii, f. 13, p. 38. *Carboniferous Limestone; Rahans Bay.*

Palæechinus intermedius, W. Keeping, Q. J. G. S. XXXII (1876), pl. iii, f. 9—11, p. 37. P. M. Duncan, A. M. N. H. ser. 6, III (1889), p. 203, woodcut x. *Carboniferous Limestone; Hook Head, co. Wexford.*

P. sphæricus [Scouler MS.], McCoy, W. Keeping, Q. J. G. S. XXXII (1876), p. 38. *Carboniferous Limestone; Hook Head.*

Pelanechinus corallinus (Wright), W. Keeping, Q. J. G. S. XXXIV (1878), pl. xxxiv, f. 1—7, p. 924. T. T. Groom, Q. J. G. S. XLIII (1887), pl. xxviii, and woodcuts, p. 703. *Coral Rag; Calne.* The second specimen was presented by T. T. Groom, Esq., B.A.

Peltastes hieroglyphica (Keeping). *Salenia hieroglyphica,* W. Keeping, Foss. Upware and Brickhill (1883), pl. vii, f. 11 *a—c,* p. 136. *Lower Greensand; Brickhill.*

P. wiltshirei (Seeley). *Salenia (Hyposalenia) wiltshirei,* H. G. Seeley, A. M. N. H. ser. 3, XVII (1866), p. 180. *Red Chalk; Hunstanton.*

P. wrighti (Desor), W. Keeping, Foss. Upware and Brickhill (1883), pl. vii, f. 10 *a, b,* p. 135. *Lower Greensand; Upware.*

Perischodomus biserialis, F. McCoy, A. M. N. H. ser. 2, III (1849), p. 254, and Contrib. Brit. Pal. (1854), p. 115. W. Keeping, Q. J. G. S. XXXII (1876), pl. iii, f. 1—5, p. 35. *Carboniferous Limestone; Hook Head, co. Wexford.*

Pseudodiadema barretti, S. P. Woodward, Organic Remains, dec. v (Mem. Geol. Survey, 1856), p. 7. (This description is in part taken from specimens in the Museum.) *Cambridge Greensand; Cambridge.* Presented by J. Carter, Esq., F.G.S.

P. carteri, S. P. Woodward, Organic Remains, dec. v (Mem. Geol. Survey, 1856), p. 9, (this description is in part taken from specimens in the Museum presented by J. Carter, Esq., F.G.S.) A. J. Jukes-Browne, Q. J. G. S. XXXI (1875), pl. xv, f. 15, 16, p. 302. *Cambridge Greensand; Cambridge.*

P. fungoideum (Seeley). *Diadema fungoideum,* H. G. Seeley, A. M. N. H. ser. 3, VIII (1861), p. 19. *Cambridge Greensand; Cambridge.*

Pseudodiadema intertuberculatum (Seeley). *Diadema inter-*
tuberculatum, H. G. Seeley, A. M. N. H. ser. 3, VIII (1861),
p. 20. *Cambridge Greensand; Cambridge.*

P. inversum, H. G. Seeley, A. M. N. H. ser. 3, VIII (1861), p. 21.
Cambridge Greensand; Cambridge.

Pygaster brevifrons, v. Pygaster semisulcatus.

Pygaster semisulcatus (Phillips). *P. brevifrons,* F. McCoy,
A. M. N. H. ser. 2, II (1848), p. 414, and Contrib. Brit. Pal.
(1854), p. 61. *Inferior Oolite; Dundry.*

P. sublœvis, v. Galeropygus agariciformis.

Rhabdocidaris thurmanni, De Loriol, var. **regens,** G. F.
Whidborne, Q. J. G. S. XXXIX (1883), pl. xix, f. 13, 13 *a, b,*
p. 536. *Inferior Oolite; Dundry.* Presented by the Rev.
G. F. Whidborne, M.A.

Rhoechinus irregularis, W. Keeping, Q. J. G. S. XXXII (1876),
pl. iii, f. 6—8, p. 37. *Carboniferous Limestone; Hook Head, co.*
Wexford.

Salenia hieroglyphica, v. Peltastes hieroglyphica.

S. (Hyposalenia) wiltshirei, v. Peltastes wiltshirei.

Salenia woodwardi, H. G. Seeley, A. M. N. H. ser. 3, VIII
(1861), p. 16. *Cambridge Greensand; Cambridge.*

Stomechinus germinans (Phillips). *Echinus diademata,* F.
McCoy, A. M. N. H. ser. 2, II (1848), p. 410, and Contrib.
Brit. Pal. (1854), p. 57. *Great Oolite; Minchinhampton.* The
specimen from the Coralline Oolite of Malton cannot be found.

S. gyratus (Agassiz). *Echinus petallatus,* F. McCoy, A. M. N. H.
ser. 2, II (1848), p. 409, and Contrib. Brit. Pal. (1854), p. 57.
T. Wright, Brit. Foss. Echin. Oolit. (1856), p. 217. *Coralline*
Oolite; Calne.

ASTEROIDEA.

Asterias (*Astropecten*) *recta*, v. Astropecten rectus.

Astropecten hastingiæ, Forbes, T Wright, Brit. Foss. Echinod. Oolit. II (1863), pl. vi, f. 4 *a, b*, p. 113. *Middle Lias* (*Marlstone*)*; Boulby, near Staithes.* Leckenby Collection. The specimen is missing.

A. rectus, McCoy. *Asterias* (*Astropecten*) *recta*, F. McCoy, A. M. N. H. ser. 2, II (1848), p. 408, and Contrib. Brit. Pal. (1854), p. 56. *Calcareous Grit; Filey Brig.*
Another specimen, T. Wright, Brit. Foss. Echin. Oolit. II (1880), pl. xix, f. 1 *a, b*, p. 168. *Lower Calcareous Grit; Scarborough.* Leckenby Collection.

A. scarburgensis, T. Wright, Brit. Foss. Echinod. Oolit. II (1863), pl. vii, f. 2 *a, b, c*, p. 115. *Inferior Oolite* (*Grey Limestone*)*; near Scarborough.* Leckenby Collection.

Goniaster (*Goniodiscus*) *rectilineus*, v. Pentagonaster rectilineus.

Palæaster hirudo (Forbes). *Uraster hirudo*, E. Forbes, Organic Remains, dec. I (Mem. Geol. Surv. 1849), pl. i, f. 4, p. 3. F. McCoy, Brit. Palæoz. Foss. (1851), p. 60. *Lower Ludlow; Potter's Fell, Kendal.*

P. ruthveni (Forbes), Salter, Cat. (1873), p. 163 and A. M. N. H. ser. 2, XX (1857), p. 326. *Uraster ruthveni*, E. Forbes, Organic Remains, dec. I (Mem. Geol. Surv. 1849), pl. i, f. 1, p. 1. F. McCoy, Brit. Palæoz. Foss. (1851), p. 59. *Lower Ludlow; High Thorns, Underbarrow.*

P. squamatus, Salter, Cat. (1873), p. 47. *Middle Bala; Bala.* Presented by the Rev. J. Peters.

Palæasterina primæva (Forbes). *Uraster primævus*, E. Forbes, Organic Remains, dec. I (Mem. Geol. Surv. 1849), pl. i, f. 2 *a, b*, p. 2. F. McCoy, Brit. Palæoz. Foss. (1851), p. 60. *Lower Ludlow; High Thorns, Underbarrow.*

P. ramseyensis, H. Hicks, Q. J. G. S. XXIX (1873), pl. iv, f. 23, p. 51. *Tremadoc Rocks; St Davids.* Presented by Dr H. Hicks, F.R.S.

Pentagonaster rectilineus (McCoy). *Goniaster (Goniodiscus) rectilineus*, F. McCoy, A. M. N. H. ser. 2, II (1848), p. 408. Contrib. Brit. Pal. (1854), p. 55. *Upper Chalk; Norwich.*

Plumaster ophiuroides, T. Wright, Brit. Foss. Echinod. Oolit. II (1863), pl. v. f. 1 *a, b*, p. 112. *Middle Lias; near Skinningrave Bay, Yorkshire.* (This is the locality given by Dr Wright, but the specimen is labelled Lower Lias, Robin Hood's Bay.) Leckenby Collection.

Uraster carinatus, T. Wright, Brit. Foss. Echinod. Oolit. II (1863), pl. ii, f. 1, p. 101. *Middle Lias (Marlstone); Boulby, Yorkshire.* Leckenby Collection.

Uraster hirudo, v. Palæaster hirudo.

U. primævus, v. Palæasterina primæva.

U. ruthveni, v. Palæaster ruthveni.

OPHIUROIDEA.

Ophioderma carinata, T. Wright, Brit. Foss. Echinod. Oolit. II (1866), pl. xvi, f. 1 *a, b*, p. 148. *Middle Lias (Marlstone); Staithes.* Leckenby Collection.

Ophiolepis leckenbyi, T. Wright, Brit. Foss. Echinod. Oolit. II (1880), pl. xix, f. 3 *a, b*, p. 160. *Inferior Oolite (Grey Limestone); near Scarborough.* Leckenby Collection. In the explanation of the plate, this specimen is erroneously stated to be in Dr Wright's collection. Fig. 3*a* is said to be of the natural size, but is magnified two diameters.

Ophiura salteri, v. Protaster salteri.

Protaster petri, Salter, Cat. (1873), p. 47. *Middle Bala; Bala.* Presented by the Rev. J. Peters.

P. salteri (Forbes), J. W. Salter, A. M. N. H. ser. 2, xx (1857), p. 332, and Appendix to A. C. Ramsay, Geol. North Wales (1866), pl. xxiii, f. 3, 3 *a*, p. 289, and *ibid.* second edition (1881), p. 480, and Cat. Camb. Sil. Foss. (1873), p. 46. *Ophiura salteri*, Q. J. G. S. I (1845), p. 20 (table). *Middle Bala; Pen-y-gair, Cerrig-y-Druidion.*

Protaster sedgwicki, E. Forbes, Organic Remains, dec. I (Mem. Geol. Surv. 1849), pl. iv, f. 1—4. F. McCoy, Brit. Palæoz. Foss. (1851), p. 60. *Lower Ludlow; Docker Park, near Kendal.*

CRINOIDEA.

Actinocrinus (Amphoracrinus) atlas, v. Amphoracrinus atlas.

A. (Amphoracrinus?) olla, v. Amphoracrinus olla.

A.? pulcher [Salter MS.], F. McCoy, Brit. Palæoz. Foss. (1851), pl. i D, f. 3, p. 55. Salter, Cat. (1873), p. 92. *Denbighshire Flags; Nant-gwrhyd-uchaf, S. of Llangollen.* (Periechocrinus?)

Actinometra loveni, P. H. Carpenter, Q. J. G. S. XXXVI (1880), p. 51, woodcut. *Gault; Folkestone.*

A. wurtembergiæ, P. H. Carpenter, Journ. Linn. Soc. Zool. XV (1881), pl. ix, f. 7, p. 197. *Weisser Jura ε (Corallian); Nattheim.*

Amphoracrinus atlas, McCoy. *Actinocrinus (Amphoracrinus) atlas,* F. McCoy, A. M. N. H. ser. 2, III (1849), p. 248, and Contrib. Brit. Pal. (1854), p. 109, and Brit. Palæoz. Foss. (1851), pl. iii D, f. 5, p. 120. *Carboniferous Limestone; Bolland.*

A. olla, McCoy. *Actinocrinus (Amphoracrinus?) olla,* F. McCoy, A. M. N. H. ser. 2, III (1849), p. 247, and Contrib. Brit. Pal. (1854), p. 109, and Brit. Palæoz. Foss. (1851), pl. iii D, f. 6, p. 121. *Carboniferous Limestone; Derbyshire.*

Antedon aspera (Quenstedt), P. H. Carpenter, Journ. Linn. Soc. Zool. XV (1881), pl. xi, f. 19 *a, b, c.* p. 202. *Weisser Jura ε (Corallian); Streitberg.*

A. costata (Goldfuss), P. H. Carpenter, Journ. Linn. Soc. Zool. XV (1881), pl. ix, f. 2, p. 191. *Weisser Jura ε (Corallian); Nattheim.*

A. depressa, P. H. Carpenter, Journ. Linn. Soc. Zool. XV (1881), pl. x, f. 12, 13, p, 201. *Weisser Jura ε (Corallian); Nattheim.*

A. scrobiculata (Goldfuss), P. H. Carpenter, Journ. Linn. Soc. Zool. XV (1881), pl. x, f. 17, 18, p. 203. *Weisser Jura ε (Corallian); Streitberg.*

Apiocrinus exutus, F. McCoy, A. M. N. H. ser. 2, II (1848), p. 406, and Contrib. Brit. Pal. (1854), p. 53. *Bradford Clay; Bradford.*

Botryocrinus decadactylus (Salter). F. A. Bather, A. M. N. H. ser. 6, VII (1891), pl. xiii, f. 13, 14, p. 395. *Cyathocrinus decadactylus*, Salter, Cat. (1873), p. 123. *Cyathocrinus quindecimalis*, Salter, Cat. (1873), p. 124. *Wenlock Limestone; Dudley.* Fletcher Collection.

Bourgueticrinus cylindricus, F. McCoy, A. M. N. H. ser. 2, II (1848), p. 404. Contrib. Brit. Pal. (1854), p. 52. *Upper Chalk; Norwich.*

B. milleri, F. McCoy, A. M. N. H. ser. 2, II (1848), p. 405. Contrib. Brit. Pal. (1854), p. 53. *Upper Chalk; Norwich.*

B. ooliticus, F. McCoy, A. M. N. H. ser. 2, II (1848), p. 405, and Contrib. Brit. Pal. (1854), p. 53. *Bradford Clay; Bradford.*

Calceocrinus abdominalis (Salter). *Cheirocrinus abdominalis*, Salter, Cat. (1873), p. 118. *Wenlock Limestone; Dudley.* Fletcher Collection. The only specimen in the Museum is the one bearing the number a/386, but Mr F. A. Bather states that this is a *Desmidocrinus*, and certainly not the type of *C. abdominalis.*

C. fletcheri (Salter). *Cheirocrinus fletcheri*, Salter, Cat. (1873), p. 119. *Wenlock Limestone; Dudley.* Fletcher Collection.

C. gradatus (Salter). *Cheirocrinus gradatus*, Salter, Cat. (1873), p. 118. *Wenlock Limestone; Dudley.* Fletcher Collection.

C. serialis [Austin MS. sp.]. *Cheirocrinus serialis*, Salter, Cat. (1873), p. 118. *Wenlock Limestone; Dudley.* Fletcher Collection. The specimens labelled *C. serialis* by Salter, belong to more than one species. F. A. Bather.

Cheirocrinus abdominalis, v. Calceocrinus abdominalis.

C. fletcheri, v. Calceocrinus fletcheri.

C. gradatus, v. Calceocrinus gradatus.

C. serialis, v. Calceocrinus serialis.

Cromyocrinus nuciformis (McCoy). *Poteriocrinus nuciformis*, F. McCoy, A. M. N. H. ser. 2, III (1849), p. 245, and Contrib. Brit. Pal. (1854), p. 106, and Brit. Palæoz. Foss. (1851), pl. iii D,

f. 4, 4 *a*, p. 117. *Carboniferous Limestone; Derbyshire.* Presented by W. Hopkins, Esq.

Cyathocrinus acinotubus, Angelin. *C. monile*, Salter, Cat. (1873), p. 124. *Wenlock Limestone; Dudley.* Fletcher Collection.

C. arboreus, Salter, Cat. (1873), p. 125. *Wenlock Limestone; Dudley.* Fletcher Collection. The specimen is missing.

C. decadactylus, v. Botryocrinus decadactylus.

C. ichthyocrinoides, Salter, Cat. (1873), p. 125. *Wenlock Limestone; Dudley.* Fletcher Collection.

C. monile, v. Cyathocrinus acinotubus.

C. nodulosus, Salter, Cat. (1873), p. 124. *Wenlock Limestone; Railway Tunnel, Dudley.* Fletcher Collection. (Taxocrinus sp.)

C. quindecimalis, v. Botryocrinus decadactylus.

C. scoparius, Salter, Cat. (1873), p. 125. *Wenlock Limestone; Dudley.* Fletcher Collection. Mr F. A. Bather states that the description ' in no way agrees with the specimen numbered a/491, which is almost certainly a *Gissocrinus.*'

C. squamiferus, Salter, Cat. (1873), p. 124. *Wenlock Limestone; Wren's Nest, Dudley.* Fletcher Collection. (Gissocrinus.)

C. sp. 1, (Salter), v. Gissocrinus goniodactylus.

C. sp. 5, (Salter), v. ? Gissocrinus goniodactylus.

C. sp. 11, Salter, Cat. (1873), p. 125. *Wenlock Limestone; Dudley.* Fletcher Collection. The specimen is missing.

C. spp. Salter, Cat. (1873), p. 124. *Wenlock Limestone; Dudley.* Fletcher Collection.

C. sp. ' *Part of the pelvis of a crinoidal animal,*' J. de C. Sowerby Trans. Geol. Soc. ser. 2, v (1840), pl. lv, f. 8. *Devonian; Looe.*

Cyclocrinus* variolarius (Seeley). *Torynocrinus? variolarius,* H. G. Seeley, A. M. N. H. ser. 3, xvii (1866), p. 174. *Koninckocrinus rugosus* (d'Orbigny), H. G. Seeley, *ibid.* xiv (1864), p. 277. *Red Chalk; Hunstanton.*

Cupressocrinus calyx, v. Hydreinocrinus calyx.

C. impressus, v. Hydreinocrinus m'coyanus.

---

* The name *Cyclocrinus* is preoccupied by Eichwald, no other has yet been proposed.

Dendrocrinus? **cambrensis**, H. Hicks, Q. J. G. S., XXIX (1873), pl. iv, f. 17—19, [20 gutta percha cast], p. 51. *Tremadoc Rocks; Ramsey Island, St Davids.* Presented by Dr H. Hicks.

Dimerocrinus multiplex, Salter, Cat. (1873), p. 120. *Wenlock Limestone; Dudley.* Fletcher Collection. (Encrinus speciosus? Angelin.)

Dimerocrinus uniformis, Salter, Cat. (1873), p. 120. *Wenlock Limestone; Dudley.* Fletcher Collection. (Carpocrinus?)

Extracrinus dichotomus (McCoy). *Pentacrinus dichotomus,* F. McCoy, A. M. N. H. ser. 2, II (1848), p. 406, and Contrib. Brit. Pal. (1854), p. 54. *Lias; Whitby.*

E. fossilis (Blumenbach). '*Briarœan Pentacrinite,*' J. Parkinson, Organic Remains II (1808), pl. xvii, f. 9, p. 251; f. 15, 16, pp. 249, 252; pl. xviii, f. 1, p. 252. *Lias; Charmouth,* except pl. xvii, f. 15, 16, from *Bath.*

    Pentacrinites briareus, W. Buckland, Bridgewater Treatise (1836), pl. liii, f. 1—5, p. 434, and *ibid.* fourth edition (1869—70), pl. lxxiii, f. 1—5, p. 362. *Lias; Lyme Regis.*

Gissocrinus goniodactylus (Phillips). *Cyathocrinus sp.* 1, Salter, Cat. (1873), p. 123. ?*C. sp.* 5, Salter, *ibid.* p. 124. *Wenlock Limestone; Dudley.* Fletcher Collection.

Glyptocrinus basalis, F. McCoy, A. M. N. H. ser. 2, VI (1850), p. 289, and Contrib. Brit. Pal. (1854), p. 177, and Brit. Palæoz. Foss. (1851), pl. i D, f. 4, p. 57. *Middle Bala; Allt-yr-Anker, Meifod, Montgomeryshire.*

G. sp. 1, Salter, Cat. (1873), p. 122. *Wenlock Limestone; Dudley.* Fletcher Collection.

Herpetocrinus fletcheri, Salter, Cat. (1873), p. 118. *Wenlock Limestone; Dudley.* Fletcher Collection.

Hydreionocrinus calyx (McCoy). *Cupressocrinus calyx,* F. McCoy, A. M. N. H. ser. 2, III (1849), p. 244, and Contrib. Brit. Pal. (1854), p. 105, and Brit. Palæoz. Foss. (1851), pl. iii D, f. 1, 1 a, p. 117. *Carboniferous Limestone; Derbyshire.*

H. m'coyanus (De Koninck and Le Hon). *Cupressocrinus impressus,* F. McCoy, A. M. N. H. ser. 2, III (1849), p. 244, and Contrib. Brit. Pal. (1854), p. 106, and Brit. Palæoz. Foss. (1851), pl. iii D, f. 2, 2 a, p. 117 (*Poteriocrinus excavatus,* foot-note to explanation of pl. iii D). *Carboniferous Limestone; Derbyshire.* Presented by W. Hopkins, Esq.

Ichthyocrinus bacchus, Salter, Cat. (1873), p. 126. *Wenlock Limestone; Wren's Nest, Dudley.* Fletcher Collection.

I. m'coyanus, Salter, Cat. (1873), p. 163. *Lower Ludlow; Light Beck, Underbarrow near Kendal.*

Koninckocrinus agassizi, v. Torynocrinus canon.

K. rugosus, v. Cyclocrinus variolarius.

Mariacrinus flabellatus, v. Melocrinus flabellatus.

Melocrinus flabellatus (Salter). *Mariacrinus flabellatus,* Salter, Cat. (1873), p. 122. *Wenlock Limestone; Dudley.* Fletcher Collection.

Millericrinus pratti (Gray). P. H. Carpenter, Q. J. G. S. XXXVIII (1882), pl. i, f. 1—8, 10—23, p. 29. '*The Lansdown Encrinite,*' H. Jelly, Bath and Bristol Mag. II (1833), p. 36, plate. *Great Oolite; Lansdown, near Bath.* Walton Collection.

Pentacrinites briareus, v. Extracrinus fossilis.

Pentacrinus dichotomus, v. Extracrinus dichotomus.

Pentacrinus goldfussi, F. McCoy, A. M. N. H. ser. 2, II (1848), p. 407, and Contrib. Brit. Pal. (1854), p. 54. *Middle Lias; Gloucestershire.*

Periechocrinus limonium, Salter, Cat. (1873), p. 121. *Wenlock Limestone; Dudley.* Fletcher Collection.

P. simplex, Salter, Cat. (1873), p. 121. *Wenlock Limestone; Dudley.* Fletcher Collection.

Platycrinus diadema, F. McCoy, A. M. N. H. ser. 2, III (1849), p. 246, and Contrib. Brit. Pal. (1854), p. 108. *Carboniferous Limestone; Cleenish, co. Fermanagh.*

P. megastylus, F. McCoy, A. M. N. H. ser. 2, III (1849), p. 247, and Contrib. Brit. Pal. (1854), p. 108, and Brit. Palæoz. Foss. (1851), p. 119. *Carboniferous Limestone; Bolland.*

P.? pecten, Salter, Cat. (1873), p. 122. *Wenlock Limestone; Wren's Nest, Dudley.* Fletcher Collection.

P. vesiculosus, F. McCoy, A. M. N. H. ser. 2, III (1849), p. 246, and Contrib. Brit. Pal. (1854), p. 107, and Brit. Palæoz. Foss.

(1851), pl. iii D, f. 3, p. 119. *Carboniferous Limestone; near Bakewell, Derbyshire.*

Poteriocrinus crassimanus, F. McCoy, A. M. N. H. ser. 2, III (1849), p. 245, and Contrib. Brit. Pal. (1854), p. 107. *Carboniferous Limestone; Hook Head, co. Wexford.* The specimen cannot be found.

P. excavatus, v. Hydreinocrinus m'coyanus.

P. nuciformis, v. Cromyocrinus nuciformis.

Taxocrinus? granulatus, Salter, Cat. (1873), p. 126. *Wenlock Limestone; Dudley.* Fletcher Collection.

T. marmoratus, v. Taxocrinus tuberculatus.

T. nanus, Salter, Cat. (1873), p. 126. *Wenlock Limestone; Wren's Nest, Dudley.* Fletcher Collection.

T.? orbignii, F. McCoy, A. M. N. H. ser. 2, VI (1850), p. 289, and Contrib. Brit. Pal. (1854), p. 176, and Brit. Palæoz. Foss. (1851), pl. i D, f. 1, p. 53. *Lower Ludlow; High Thorns, Underbarrow.*

T. tuberculatus (Miller). *T. marmoratus*, Salter, Cat. (1873), p. 125. *Wenlock Limestone; Dudley.* Fletcher Collection.

Torynocrinus canon, H. G. Seeley, A. M. N. H. ser. 3, XVII (1866), p. 174. *Koninckocrinus agassizi*, H. G. Seeley, A. M. N. H. ser. 3, XIV (1864), p. 277. *Red Chalk; Hunstanton.*

T.? variolarius, v. Cyclocrinus variolarius.

'*Crinoidal remains*,' J. de C. Sowerby, Trans. Geol. Soc. ser. 2, V (1840), pl. liii, f. 17—21. *Devonian; fig. 17, Barnstaple, f. 18, 19, 21, Bedruthen Steps, St Eval, f. 20, Lower St Columb Porth.*

'*Crinoidal casts*,' J. de C. Sowerby, Trans. Geol. Soc. ser. 2, V (1840), pl. liii, f. 31. *Upper Devonian; Croyde Bay.*

CYSTIDEA.

Ateleocystites fletcheri, v. Atelecystis forbesi.

Atelecystis forbesi (De Koninck). *Ateleocystites fletcheri*, Salter, Cat. (1873), p. 128. *Wenlock Limestone; Wren's Nest, Dudley.* Fletcher Collection.

Caryocystites davisi, v. Echinosphæra davisi.

Echino-encrinus armatus, E. Forbes, Mem. Geol. Survey, II part ii (1848), pl. xviii, p. 507. *Wenlock Shale; Walsall and Malvern.* The description and figures of this species are taken in part from the specimens in the Museum.

Echinosphæra davisi (McCoy). *Carocystites davisi*, F. McCoy, Brit. Palæoz. Foss. (1855), pl. i D, f. 5, 5 *a*, [6 ? missing], p. 61. *Coniston Limestone; Coniston.*

Lepadocrinus bifasciatus (Pearce). *Pseudocrinites bifasciatus*, E. Forbes, Mem. Geol. Survey II, part ii (1848), pl. xi, p. 496. *Wenlock Limestone; Dudley.* Fletcher Collection. The descriptions and figures given by Forbes of this and the next species, are taken in part from the specimens in the Museum.

L. quadrifasciatus (Pearce). *Pseudocrinites quadrifasciatus*, E. Forbes, Mem. Geol. Survey II, part ii (1848), pl. xiii, p. 48. *Wenlock Limestone; Dudley.* Fletcher Collection. See note to the preceding species.

Protocystis meneviensis, H. Hicks, Q. J. G. S. XXVIII (1872), pl. v, f. 19, p. 180. *Menevian Group; Porth-y-rhaw, St Davids.* Presented by Dr H. Hicks, F.R.S.

Prunocystis fletcheri, E. Forbes, Mem. Geol. Survey II, part ii (1848), pl. xvi, f. 1—4, p. 503. *Wenlock Limestone; Dudley.* Fletcher Collection.

Pseudocrinites bifasciatus, v. Lepadocrinus bifasciatus.

P. quadrifasciatus, v. Lepadocrinus quadrifasciatus.

BLASTOIDEA.

Codonaster acutus, v. Codonaster trilobatus *var.* acutus.

Codonaster trilobatus, [*Codaster*, McCoy], F. McCoy, A. M. N. H. ser. 2, III (1849), p. 251, and Contrib. Brit. Pal. (1854), p. 112, and Brit. Palæoz. Foss. (1851), pl. iii D, f. 8, 8 *a*, p. 123. *Carboniferous Limestone; Derbyshire.* Presented by W. Hopkins, Esq.

C. trilobatus, var. **acutus**, McCoy. *C. acutus*, F. McCoy, A. M N. H. ser. 2, III (1849), p. 251, and Contrib. Brit. Pal. (1854)

44 BLASTOIDEA.

p. 112, and Brit. Palæoz. Foss. (1851), pl. iii D, f. 7, 123. *Carboniferous Limestone; Bolland, Yorkshire.*

Granatocrinus campanulatus (McCoy). *Pentremites campanulatus,* F. McCoy, A. M. N. H. ser. 2, III (1849), p. 249, and Contrib. Brit. Pal. (1854), p. 111, and Brit. Palæoz. Foss. (1851), pl. iii D, f. 9, p. 123. *Carboniferous Limestone; Derbyshire.* Presented by W. Hopkins, Esq.

Pentremites campanulatus, v. Granatocrinus campanulatus.

VERMES.

Crossopodia lata, F. McCoy, A. M. N. H. ser. 2, VII (1851), p. 395, and Contrib. Brit. Pal. (1854), p. 130, and Brit. Palæoz. Foss. (1851), pl. i D, f. 14, p. 130. *Upper Ludlow Tilestones; Storm Hill, Llandeilo.*

C. scotica, F. McCoy, A. M. N. H. ser. 2, VII (1851), p. 395, and Contrib. Brit. Pal. (1854), p. 179, and Brit. Palæoz. Foss. (1851), pl. i D, f. 15, p. 130. *Bala Beds; Thornilee Quarry, near the river Tweed, Selkirkshire.*

Nereites cambrensis, R. I. Murchison, Sil. Syst. (1839), pl. xxvii, f. 1, p. 700, and Siluria, ed. 5 (1872), p. 201, woodcut. *May Hill Group; Lampeter.*

Palæochorda major, F. McCoy, Q. J. G. S. IV (1848), p. 225, and Brit. Palæoz. Foss. (1851), pl. i A, f. 3. *Skiddaw Slates; Kirkfell, near Scawgill.*

P. minor, F. McCoy, Q. J. G. S. IV (1848), p. 225, and Brit. Palæoz. Foss. (1851), pl. i A, f. 1, 2. *Skiddaw Slates; fig. 1 from Blakefell, f. 2 from Under Crag.*

Serpula articulata, Sowerby, W. Keeping, Foss. Upware and Brickhill (1883), pl. vii, f. 7, p. 131. *Lower Greensand; Upware.*

S. lophioda, Goldfuss, W. Keeping, Foss. Upware and Brickhill (1883), pl. vii, f. 5 a, b, p. 131. *Lower Greensand; Upware.*

S. rustica, Sowerby, W. Keeping, Foss. Upware and Brickhill (1883), pl. vii, f. 6 a, b, p. 131. *Lower Greensand; Upware.*

S. sp.?, W. Keeping, Foss. Upware and Brickhill (1883), p. 132. *Lower Greensand; Brickhill.*

Serpulites dispar, J. W. Salter, Appendix to F. McCoy, Brit. Palæoz. Foss. (1855), p. i. F. McCoy, *ibid.* pl. i D, f. 11, 12, p. 132. *Upper Ludlow; Benson Knot.*

S.? longissimus, v. Trachyderma lævis.

Trachyderma lævis, F. McCoy, A. M. N. H. ser. 2, VII (1851), p. 396, and Contrib. Brit. Pal. (1854), p. 179, and Brit. Palæoz. Foss. (1851), pl. i D, f. 10, p. 133. *Serpulites? longissimus,* Salter, Cat. (1873), p. 47. *Caradoc Beds; Acton Scott.*

Vermicularia contorta, J. F. Blake, Q. J. G. S. XXXI (1875) p. 231. *Kimeridge Clay; Ely.*

V. polygonalis (Sowerby), W. Keeping, Foss. Upware and Brickhill (1883), pl. vii, f. 8 *a, b*, p. 133. *Lower Greensand; Upware.*

MOLLUSCOIDEA.

POLYZOA.

Alveolites? seriatoporoides, H. M. Edwards and J. Haime, Brit. Foss. Corals (1854), pl. lxii, f. 2, 2 *a*, p. 263. *Wenlock Limestone; Dudley.* Fletcher Collection. (? Spiropora.)

Apseudesia cristata, J. Haime, Mem. Soc. Géol. Fran. ser. 2, V (1854), pl. vii, f. 6 *a, g, h*, p. 201. *Great Oolite; Hampton Cliffs, near Bath.* Walton Collection.

Berenicea contracta, v. Diastopora contracta.

B. heterogyra, F. McCoy, A. M. N. H. ser. 2, VI (1850), p. 286, and Contrib. Brit. Pal. (1854), p. 173, and Brit. Palæoz. Foss. (1855), pl. i C, f. 17, 17 *a, b*, p. 45. G. R. Vine, Q. J. G. S. XXXVI (1880), p. 358. *Coniston Limestone; Coniston.*

Cellulipora sulcata, v. Diastopora sulcata.

Ceriopora (Reptonodicava) nodosa, W. Keeping, Foss. Upware and Brickhill (1883), pl. vii, f. 14 *a, b*, p. 140. *Lower Greensand; Upware.*

Ceriopora (Echinocava) raulini, Michelin, W. Keeping, Foss. Upware and Brickhill (1883), p. 139. *Lower Greensand; Upware.*

Chrysaora similis, F. McCoy, A. M. N. H. ser. 2, II (1848), p. 404, and Contrib. Brit. Pal. (1854), p. 52. *Great Oolite; Minchinhampton.*

Diastopora contracta (Seeley), G. R. Vine, Proc. Yorks. Geol. Polytech. Soc. XI (1890), pp. 366, 378—9. *Berenicea contracta*, H. G. Seeley, A. M. N. H. ser. 3, XVII (1866), p. 181. *Red Chalk; Hunstanton.* The specimen cannot be found.

D. davidsoni, J. Haime, Mem. Soc. Géol. Fran. ser. 2, V (1854), pl. viii, f. 9 *a, b,* p. 185. *Great Oolite; Hampton Cliffs, near Bath.* Walton Collection.

D. michelini, J. Haime, Mem. Soc. Géol. Fran. ser. 2, V (1854), pl. viii, f. 8 *a, b,* p. 188. *Inferior Oolite; Charlcombe.* Walton Collection.

D. scobinula, Michelin, J. Haime, Mem. Soc. Géol. Fran. ser. 2, V (1854), pl. viii, f. 6 *a,* p. 186. *Great Oolite; Hampton Cliffs, near Bath.* Walton Collection.

D. sulcata (Seeley), G. R. Vine, Proc. Yorks. Geol. Polytech. Soc. XI (1890), pp. 378, 381. *Cellulipora sulcata*, H. G. Seeley, A. M. N. H. ser. 3, XVII (1866), p. 181. *Red Chalk; Hunstanton.*

D. waltoni, J. Haime, Mem. Soc. Géol. Fran. ser. 2, V (1854), pl. viii, f. 2 *a, b,* p. 184. *Inferior Oolite; Postlip.* Walton Collection.

D. wrighti, J. Haime, Mem. Soc. Géol. Fran. ser. 2, V (1854), pl. viii, f. 5 *a, b,* p. 186. *Inferior Oolite; Postlip.* Walton Collection.

Entalophora dendroidea, W. Keeping, Foss. Upware and Brick-hill (1883), pl. vii, f. 12 *a, b,* p. 138. *Lower Greensand; Upware.*

Fasciculipora waltoni, J. Haime, Mem. Soc. Géol. Fran. ser. 2, V (1854), pl. x, f. 4 *a—b,* p. 200. *Great Oolite; Hampton Cliffs, near Bath.* Walton Collection.

Fenestella carinata, v. F. plebeia.

Fenestella lineata, G. W. Shrubsole, Q. J. G. S. XXXVI (1880), pl. xi, f. 2, 2 *a*, p. 249. *Wenlock Limestone; Dudley.* Fletcher Collection.

F. patula, F. McCoy, A. M. N. H. ser. 2, VI (1850), p. 288, and Contrib. Brit. Pal. (1854), p. 175, and Brit. Palæoz. Foss. (1851), pl. i c, f. 20, 20 *a*, p. 50. The young stage of another species, G. W. Shrubsole, Q. J. G. S. XXXVI (1880), p. 242. *Wenlock Limestone; Dudley.*

F. plebeia, McCoy. *F. carinata*, McCoy, G. W. Shrubsole, Q. J. G. S. XXXV (1879), p. 279. *Carboniferous Limestone; Derbyshire, and Isle of Man.*

F. reteporata, G. W. Shrubsole, Q. J. G. S. XXXVI (1880), pl. xi, f. 1, 1 *a—c*, p. 249. *Wenlock Limestone; Dudley.* Fletcher Collection.

F. rigidula, F. McCoy, A. M. N. H. ser. 2, VI (1850), p. 288, and Contrib. Brit. Pal. (1854), p. 175, and Brit. Palæoz. Foss. (1855), pl. i c, f. 19, 19 *a*, p. 50. G. W. Shrubsole, Q. J. G. S. XXXVI (1880), p. 248. *Wenlock Limestone; Dudley.* The specimen is missing.

Heteropora (Nodicrescis) annulata, W. Keeping, Foss. Upware and Brickhill (1883), pl. vii, f. 16 *a*, *b*, p. 142. *Lower Greensand; Upware.*

H. (Multicresis) arbuscula, W. Keeping, Foss. Upware and Brickhill (1883), pl. vii, f. 17 *a*, *b*, p. 143. *Lower Greensand; Upware.*

H. conifera, J. Haime, Mem. Soc. Géol. Fran. ser. 2, v (1854), pl. xi, f. 1 *a*, *b*, p. 208. *Great Oolite; Hampton Cliffs, near Bath.* Walton Collection.

H. major, W. Keeping, Foss. Upware and Brickhill (1883), pl. vii, f. 18 *a*, *b*, p. 144. *Lower Greensand; Upware.*

H. (Multizonopora) ramosa, Roemer, W. Keeping, Foss. Upware and Brickhill (1883), p. 141. *Lower Greensand; Upware.*

H. (Multicrescis) sp. (allied to *H. cryptopora*, Goldfuss), W. Keeping, Foss. Upware and Brickhill (1883), p. 142. *Lower Greensand; Upware.*

Heteropora (Multicrescis) sp. (allied to *H. digitata*, Michelin), W. Keeping, Foss. Upware and Brickhill (1883), p. 142. *Lower Greensand; Upware.*

H. (Reptonodicrescis) sp., W. Keeping, Foss. Upware and Brickhill (1883), p. 143. *Lower Greensand; Upware.*

Idmonea triquetra, J. Haime, Mem. Soc. Géol. Fran. ser. 2, v (1854), pl. vii, f. 1 *a, b,* p. 171. *Bradford Clay; near Bath.* Walton Collection.

Lichenopora phillipsi, J. Haime, Mem. Soc. Géol. Fran. ser. 2, v (1854), pl. x, f. 10 *a, b,* p. 206. *Great Oolite; Hampton Cliffs, near Bath.* Walton Collection.

Melicertites upwarensis, W. Keeping, Foss. Upware and Brickhill (1883), pl. vii, f. 13 *a, b,* p. 138. *Lower Greensand; Upware.*

Neuropora damicornis, J. Haime, Mem. Soc. Géol. Fran. ser. 2, v (1854), pl. x, f. 8 *a,* p. 214. *Great Oolite; Hampton Cliffs, near Bath.* Walton Collection.

N. defrancei, J. Haime, Mem. Soc. Géol. Fran. ser. 2, v (1854), pl. x, f. 7 *a—c,* p. 215. *Great Oolite; Hampton Cliffs, near Bath.* Walton Collection.

N. spinosa, J. Haime, Mem. Soc. Géol. Fran. ser. 2, v (1854), pl. x, f. 9 *a,* p. 214. *Great Oolite; Hampton Cliffs, near Bath.* Walton Collection.

Phyllopora hisingeri (McCoy). *Retepora hisingeri,* F. McCoy, A. M. N. H. ser. 2, vi (1850), p. 477, and Contrib. Brit. Pal. (1854), p. 176, and Brit. Palæoz. Foss. (1855), pl. i c, f. 18, 18 *a,* p. 48. *Coniston Limestone; Coniston Water Head.*

Proboscina davidsoni, J. Haime, Mem. Soc. Géol. Fran. ser. 2, v (1854), pl. vi, f. 11 *a, b,* p. 167. *Great Oolite; Hampton Cliffs, near Bath.* Walton Collection.

Ptilodictya costellata, F. McCoy, A. M. N. H. ser. 2, vi (1850), p. 287, and Contrib. Brit. Pal. (1854), p. 174, and Brit. Palæoz. Foss. (1855), pl. i c, f. 15, 15 *a, b,* p. 46. *Middle Bala; Llansaintffraid.*

P. explanata, F. McCoy, A. M. N. H. ser. 2, vi (1850), p. 286, and Contrib. Brit. Pal. (1854), p. 173, and Brit. Palæoz. Foss. (1851), pl. i c, 16, 16 *a, b*, p. 46. *Middle Bala; Llansaintffraid.*

Ptilodictya fucoides, F. McCoy, A. M. N. H. ser. 2, vi (1850), p. 288, and Contrib. Brit. Pal. (1854), p. 174, and Brit. Palæoz. Foss. (1851), pl. i c, f. 14, 14 *a*, p. 47. *Middle Bala; Llansaintffraid.*

Radiopora bulbosa, d'Orbigny, **var.**, W. Keeping, Foss. Upware and Brickhill (1883), p. 139. *Lower Greensand; Upware.*

Reptocea lobosa, W. Keeping, Foss. Upware and Brickhill (1883), pl. vii, f. 15 *a, b*, p. 141. *Lower Greensand; Brickhill.*

Reptomulticava favus, H. G. Seeley, A. M. N. H. ser. 3, xvii (1866), p. 181. *Red Chalk; Hunstanton.*

Reptomultisparsa haimeana, De Loriol, W. Keeping, Foss. Upware and Brickhill (1883), p. 137. *Lower Greensand; Upware.*

Retepora hisingeri, v. Phyllopora hisingeri.

Stomatopora dichotomoides, J. Haime, Mem. Soc. Géol. Fran. ser. 2, v (1854), pl. vi, f. 2, p. 163. *Inferior Oolite; Postlip.* Walton Collection.

S. waltoni, J. Haime, Mem. Soc. Géol. Fran. ser. 2, v (1854), pl. vi, f. 3 *a, b*, p. 162. *Bradford Clay; Bradford.*

Thamniscus crassus (Lonsdale), G. W. Shrubsole, Q. J. G. S. xxxviii (1882), pp. 341, 345. *Ceriopora?*, Salter, Cat. (1873), p. 100. *Wenlock Limestone; Dudley.* Fletcher Collection.

Theonoa bowerbanki, J. Haime, Mem. Soc. Géol. Fran. ser. 2, v (1854), pl. x, f. 3, p. 205. *Inferior Oolite; Postlip, near Cheltenham.* Walton Collection.

BRACHIOPODA.

Athyris concentrica (Von Buch), F. McCoy, Brit. Palæoz. Foss. (1852), p. 378. *Atrypa hispida*, J. de C. Sowerby, Trans. Geol. Soc. ser. 2, v (1840), pl. liv, f. 4. *A. decussata*, J. de C. Sowerby, *ibid.* (1840), pl. liv, f. 5. T. Davidson, Brit. Foss.

Brach. III, part vi, (1864), pl. iii, f. 17, p. 17.   *Upper Devonian;*
South Devon.

Athyris expansa (Phillips), T. Davidson, Brit. Foss. Brach. II,
part V, (1861), pl. xvii, f. 3, 4, p. 82.   *Carboniferous Limestone;*
Settle.   Burrow Collection.

A. globularis (Phillips), T. Davidson, Brit. Foss. Brach. II,
part V (1861), pl. xvii, f. 16, 16 *a*, *b*, 17, 17 *a*, 18, p. 87.   *Carbo-*
niferous Limestone; Settle.   Burrow Collection.

A. gregaria var. *trapezoidalis*, v. Athyris subtilita.

A. lamellosa (L'Eveillé), T. Davidson, Brit. Foss. Brach. II,
part V (1861), pl. xvii, f. 6, 6 *a*, 7, p. 79.   *Carboniferous*
Limestone; Settle.   Burrow Collection.

A. obovata, v. Glassia obovata.

A. subtilita (Hall).   *A. gregaria*, McCoy, var. *trapezoidalis*, F.
McCoy, Brit. Palæoz. Foss. (1855), pl. iii D, f. 20*, 20* *a*, *b*, *c*,
p. 435.   *Carboniferous Limestone; Kendal.*

Atrypa cuboides, v. Rhynchonella cuboides.

A. decussata, v. Athyris concentrica.

Atrypa desquamata, J. de C. Sowerby, Trans. Geol. Soc. ser. 2,
V (1840), pl. lvi, f. 19.   *A. desquamata* var. *compressa*, J. de C.
Sowerby, *ibid.* (1840), pl. lvi, f. 21.   *Middle Devonian; Ply-*
mouth.

A. fallax, v. Rhynchonella pleurodon.

A. flexuosa, J. E. Marr and H. A. Nicholson, Q. J. G. S. XLIV
(1888), pl. xvi, f. 20, 20 *a*, *b*, p. 725.   *Lower Skelgill Beds* (zone
of *A. flexuosa*); *Skelgill.*   Presented by J. E. Marr, Esq., M.A.

A. hispida, v. Athyris concentrica.

A.? scotica (McCoy), T. Davidson, Brit. Foss. Brach. III, part
vii (1867), pl. xiii, f. 31, 31 *a*, *b*, p. 140.   *Hemithyris hemi-*
sphærica (Sowerby) var. *scotica*, F. McCoy, Brit. Palæoz. Foss.
(1852), pl. i H, f. 10, 10 *a*, *b*, p. 202.   *Lower Llandovery; Mullock*
Quarry, near Girvan.

A. striatula, v. Orthis interlineata.

A. subdentata, v. Rhynchonella subdentata.

Atrypa triangularis, v. Rhynchonella acuminata *var.* mesogona.

A. unguiculus, v. Spirifera urei.

Camarophoria globulina (Phillips), T. Davidson, Brit. Foss. Brach. II, part v (1861), pl. xxiv, f. 17, p. 115. *Hemithyris longa*, F. McCoy, A. M. N. H. ser. 2, x (1852), p. 425, and Contrib. Brit. Pal. (1854), p. 252, and Brit. Palæoz. Foss. (1855), pl. iii D, f. 24, 24 *a—c*, p. 440. *Carboniferous Limestone; Derbyshire.* Presented by W. Hopkins, Esq.

C.? laticliva, F. McCoy, Brit. Palæoz. Foss. (1855), pl. iii D, f. 20 *a—c*, 21, p. 444. T. Davidson, Brit. Foss. Brach. II, part iii (1861), pl. xxv, f. 11, *a, b,* 12, p. 116. *Carboniferous Limestone.* Figs. 20 (McCoy) and 11 (Davidson) from *Derbyshire,* presented by W. Hopkins, Esq. Figs. 21 and 12 from *Lowick.*

Chonetes buchiana, De Koninck. *Leptœna (Chonetes) crassistria,* F. McCoy, Brit. Palæoz. Foss. (1855), pl. iii H, f. 5, 5 *a,* p. 454. *Carboniferous Limestone; Derbyshire.* Presented by W. Hopkins, Esq.

C. hardrensis, Phillips, T. Davidson, Brit. Foss. Brach. II, part v (1863), pl. xlvii, f. 24, p. 188. *Leptœna (Chonetes) subminima,* F. McCoy, A. M. N. H. ser. 2, x (1852), p. 428, and Contrib. Brit. Pal. (1854), p. 255, and Brit. Palæoz. Foss. (1855), pl. iii D, f. 31, p. 456. *Carboniferous Limestone; Derbyshire.*

C. polita, McCoy, T. Davidson, Brit. Foss. Brach. II, part v (1863), pl. xlvii, f. 10, p. 190. *Leptœna (Chonetes) polita,* F. McCoy, A. M. N. H. ser. 2, x (1852), p. 427, and Contrib. Brit. Pal. (1854), p. 254, and Brit. Palæoz. Foss. (1855), pl. iii D, f. 30, p. 456. *Carboniferous Limestone; Lowick.*

Crania antiquior [Jelly MS.], T. Davidson, London Geol. Journal, part iii (1847), pl. xviii, f. 21—25, p. 115, and Brit. Foss. Brach. I, part iii (1851), pl. i, f. 4—8, p. 11. *Great Oolite; Hampton Cliffs, near Bath.* Walton Collection.

C. divaricata (McCoy), T. Davidson, Brit. Foss. Brach. III, part vii (1866), pl. viii, f. 7, 8, p. 78. *Pseudocrania divaricata,* F. McCoy, A. M. N. H. ser. 2, VIII (1851), p. 388, and Contrib. Brit. Palæont. (1854), p. 210, and Brit. Palæoz. Foss. (1852), pl. i H, f. 1, 1 *a,* 2, 2 *a,* p. 187. *Middle Bala; Bryn-melyn, near Bala.*

Cyrtina carbonaria (McCoy), T. Davidson, Brit. Foss. Brach.
II, part v (1859), pl. xv, f. 5, 13, 14, p. 71. *Pentamerus carbonarius*, F. McCoy, A. M. N. H. ser. 2, x (1852), p. 426, and
Contrib. Brit. Pal. (1854), p. 253, and Brit. Palæoz. Foss. (1855),
pl. iii D, f. 12—18, p. 442. *Carboniferous Limestone; Kendal.*

C. septosa (Phillips), T. Davidson, Brit. Foss. Brach. II, part v
(1863), pl. l, f. 19, 19 *a*. *Carboniferous Limestone; Settle.*
Burrow Collection.

Dinobolus? **hicksi** [Davidson MS.], H. Hicks, Q. J. G. S. XXXI
(1875), pl. x, f. 6, p. 188. *Llanvirn Beds; Llanvirn Quarry.*
Presented by Dr H. Hicks, F.R.S.

Discina bulla, v. D. nitida.

Discina (Trematis) corona, Salter, Cat. (1873), p. 55. T.
Davidson, Brit. Foss. Brach. III, part vii (1871), pl. xlix, f. 43,
44, p. 344. *Middle Bala; Pusgill, Dufton.*

D. nitida (Phillips). *D. bulla*, F. McCoy, A. M. N. H. ser. 2,
x (1852), p. 421, and Contrib. Brit. Pal. (1854), p. 248,
and Brit. Palæoz. Foss. (1855), pl. iii D, f. 32, 32 *a*, p. 407.
Carboniferous Limestone; Lowick.

D. pileolus [Hicks MS.], T. Davidson, G. M. v (1868), pl. xvi, f. 13,
13 *a*, p. 312. *D. pileolulus?*, H. Hicks, Q. J. G. S. XXVII
(1871), pl. xv, f. 12, 12 *a*. *Harlech Beds; St Davids.* Presented by Dr H. Hicks.

Glassia obovata (Sowerby). *Athyris obovata*, Salter, Cat.
(1873), p. 139. *Spondylobolus craniolaris*, F. McCoy, A. M.
N. H. ser. 2, VIII (1851), p. 408, and Contrib. Brit. Pal. (1854),
p. 230, and Brit. Palæoz. Foss. (1852), pl. i H, f. 4—5, p. 255.
T. Davidson, Brit. Foss. Brach. III, part vii (1866), pl. viii,
f. 26, 27. *Wenlock Shale; Builth Bridge.*

Hemithyris acuminata var. *mesogona*, v. Rhynchonella acuminata
var. mesogona.

H. angustifrons, v. Meristella angustifrons.

H. davidsoni, v. Rhynchonella wilsoni *var*. davidsoni.

H. hemisphœrica var. *scotica*, v. Atrypa? scotica.

H. heteroptycha, v. Rhynchonella flexistria.

Hemithyris longa, v. Camarophoria globulina.

H. nasuta, v. Rhynchonella nasuta.

H. sphæroidalis, v. Rhynchonella wilsoni *var.* sphæroidalis.

H. subdentata, v. Rhynchonella subdentata.

H. subundata, v. Meristella? subundata.

Kingena? **rhomboidalis**, Keeping, T. Davidson, Brit. Foss. Brach. v (1884), pl. xviii, f. 10, 10 *a*, 10 *b*, p. 246. *Megerlia (Kingena) rhomboidalis*, W. Keeping, Foss. Upware and Brickhill (1883), pl. vii, f. 3 *a—c*, p. 128. *Lower Greensand; Brickhill.*

Koninckella (Leptæna?) **liasiana** (Bouchard), T. Davidson, Brit. Foss. Brach. v (1884), pl. xx, f. 18, 18 *a*, *b*, p. 278. *Middle Lias (Cymbium-zone); Chideock, Dorset.* Presented by the Rev. G. F. Whidborne, M.A.

Leptæna crassistria, v. Chonetes buchiana.

L. deltoida var. *undata*, v. Strophomena deltoidea *var.* undata.

L. fragaria var., v. Productus subaculeatus.

L. gigas, v. Streptorhynchus gigas.

Leptæna? **nobilis** (McCoy). *Strophomena nobilis*, F. McCoy, A. M. N. H. ser. 2, VIII (1851), p. 486, and Contrib. Brit. Pal. (1854), p. 237, and [*Leptæna*] Brit. Palæoz. Foss. (1852), pl. ii A, f. 8, 8 *a*, p. 386. *Middle Devonian; Teignmouth.*

L. polita, v. Chonetes polita.

L. prælonga, v. Productus prælongus.

L. quinquecostata (McCoy), F. McCoy, Brit. Pal. Foss. (1852), pl. i H, f. 30, 31, 31 *a—b*, 32, p. 236. *May Hill Group; Mathrafal*, (except f. 32, from *Colmonell, Stinchar*). According to Salter (Cat. Camb. Sil. Foss. p. 64), figs. 31 and 32 are *L. transversalis* var. *undulata*, Salter.

L. sericea, Sowerby, var. **rhombica**, F. McCoy, Brit. Palæoz. Foss. (1852), p. 239. *Bala Beds; Mathrafal, Allt-yr-Anker, etc.*

L. simulans, v. Strophomena simulans.

L. spiriferoides, v. Orthis? spiriferoides.

L. subminima, v. Chonetes hardrensis.

Leptæna tenuicincta (McCoy), F. McCoy, A. M. N. H. ser. 2, VIII (1851), p. 401, and Contrib. Brit. Pal. (1854), p. 223, and Brit. Palæoz. Foss. (1852), pl. i H, f. 40, 40 *a—c*, p. 239. *Bala Beds; Cefn-grugos, Llanfyllin.*

L. tenuissime-striata, F. McCoy, Brit. Palæoz. Foss. (1852), pl. i H, f. 44, 44 *a—c*, p. 239. T. Davidson, Brit. Foss. Brach. III, part vii (1871), pl. xlix, f. 22, p. 330. *L. sericea*, Sowerby, Salter, Cat. (1873), pp. 36, 64. *Llandeilo Beds; Llandeilo.*

L. (*Leptagonia*) *ungula*, v. Strophomena ungula.

L. sp., Salter, Cat. (1873), p. 36. *L. quinquecostata*, 'large depressed variety,' F. McCoy, Brit. Palæoz. Foss. (1852), pl. i H, f. 32, p. 236. *Lower Bala; Colmonell, Girvan.*

Lingula beani?, Phillips, T. Davidson, Brit. Foss. Brach. IV (1876), pl. ix, f. 14, p. 78. *Dogger Sands; Blue Wyke.* Leckenby Collection.

L. curta?, Conrad, F. McCoy, Brit. Palæoz. Foss. (1852), p. 251. T. Davidson, Brit. Foss. Brach. III, part vii (1866), pl. iii, f. 33, p. 52. *Llandeilo Beds; Wellfield, near Builth.*

L. davisi, v. Lingulella davisi.

L. latior, F. McCoy, A. M. N. H. ser. 2, x (1852), p. 429, and Contrib. Brit. Pal. (1854), p. 256, and Brit. Palæoz. Foss. (1855), pl. iii D, f. 33, p. 475. T. Davidson, Brit. Foss. Brach. II, part v (1863), pl. xlviii, f. 37, p. 210. *Carboniferous Limestone; Derbyshire.*

L.? lesueuri, Rouault, T. Davidson, Brit. Foss. Brach. IV (1881), pl. xl, f. 17, 19, p. 361. *Fig. 17 from the Budleigh Salterton Pebble bed*, presented by W. Vicary, Esq. *Fig. 19 from the Drift, Ladypool Lane, Birmingham*, presented by F. T. S. Houghton, Esq.

L. longissima?, Pander, F. McCoy, Brit. Palæoz. Foss. (1852), p. 253. T. Davidson, Brit. Foss. Brach. III, part vii (1866), pl. iii, f. 30, p. 51. *Bala Limestone; Mynydd Fron Frys, west of Chirk.*

L. obtusa?, Hall, F. McCoy, Brit. Palæoz. Foss. (1852), p. 253. T. Davidson, Brit. Foss. Brach. III, part vii (1866), pl. iii, f. 31, p. 52. *Llandeilo Beds; Llandeilo.*

Lingula ovata, v. Lingulella davisi.

L. tenuigranulata, F. McCoy, A. M. N. H. ser. 2, VIII (1851), p. 406, and Contrib. Brit. Pal. (1854), p. 228, and Brit. Palæoz. Foss. (1852), pl. i L, f. 8, 8 *a—d*, p. 254. T. Davidson, Brit. Foss. Brach. III, part vii (1866), pl. ii, f. 9, p. 37. *Middle Bala; Allt-yr-Anker, Meifod.*

Lingulella davisi (McCoy). *Lingula davisi*, F. McCoy, A. M. N. H. ser. 2, VIII (1851), p. 405, and Contrib. Brit. Pal. (1854), p. 227, and Brit. Palæoz. Foss. (1852), pl. i L, f. 7, 7 *a, b*, p. 252. *L. ovata*, F. McCoy (*pars*), Brit. Palæoz. Foss. (1852), pl. i L, f. 6, 6 *a*, p. 254. T. Davidson, Brit. Foss. Brach. III, part vii (1866), p. 39. *Tellinomya lingulæcomes*, F. McCoy, A. M. N. H. ser. 2, VII (1851), p. 56, and Contrib. Brit. Pal. (1854), p. 191, and Brit. Palæoz. Foss. (1852), pl. i K, f. 18, p. 274. *Lingula Flags; Penmorfa, Carnarvonshire.*

L. ferruginea, Salter, T. Davidson, G. M. V (1868), pl. xv, f. 1, 1 *a*, p. 306. *Harlech Beds; St Davids.* Presented by Dr H. Hicks, F.R.S.

L. primæva, H. Hicks, Q. J. G. S. XXVII (1871), pl. xv, f. 13, 14, p. 401. T. Davidson, Brit. Foss. Brach. V (1883), pl. xvii, f. 33, p. 208. *Harlech Beds; St Davids.* Presented by Dr H. Hicks.

Megerlia (Kingena) rhomboidalis, v. Kingena? rhomboidalis.

Meristella angustifrons (McCoy), T. Davidson, Brit. Foss. Brach. III, part vii (1866), pl. x, f. 21, p. 111. *Hemithyris angustifrons*, F. McCoy, A. M. N. H. ser. 2, VIII (1851), p. 391, and Contrib. Brit. Pal. (1854), p. 213, and Brit. Palæoz. Foss. (1852), pl. i H, f. 6—8, p. 199. *May Hill Group; Mullock Quarry, Girvan.*

M.? subundata (McCoy), T. Davidson, Brit. Foss. Brach. III, part vii (1867), pl. xiii, f. 4, 4 *a—c*, p. 120. *Hemithyris subundata*, F. McCoy, A. M. N. H. ser. 2, VIII (1851), p. 394, and Contrib. Brit. Pal. (1854), p. 216, and Brit. Palæoz. Foss. (1852), pl. i H, f. 9, 9 *a—c*, p. 207. *Lower Llandovery; Mathrafal, Montgomeryshire.*

Obolella maculata [Hicks MS.], T. Davidson, G. M. V (1868),

pl. xvi, f. 3, 3 *a*, p. 311, and Brit. Foss. Brach. III, part vii
(1871), pl. l, f. 21, 21 *a*, p. 341. *Menevian Beds; St Davids.*

Obolus ? **plumbeus** (Salter) var. **plicatus** [Hicks MS.], T. Davidson, Brit. Foss. Brach. III, part vii (1871), pl. l, f. 22, 22 *a*,
p. 342. *Tremadoc Beds; Tremaenhir, St Davids.* The
specimen does not agree well with the figure.

Orthis alternata, Sowerby, T. Davidson, Brit. Foss. Brach. v
(1883), pl. xiv, f. 2—6, p. 187. *Middle Caradoc; Horderley.*

O. alternata var. *retrorsistria*, v. Orthis retrorsistria.

O. budleighensis, T. Davidson, Brit. Foss. Brach. IV (1881),
pl. xli, f. 20, 20 *a*, p. 358. *Budleigh Salterton Pebble Bed.*
This specimen is missing.

O. calligramma, Dalman, var. **scotica** (McCoy), T. Davidson,
Brit. Foss. Brach. III, part vii (1869), pl. xxxv, f. 20, 20 *a, b*,
p. 240. *Orthisina scotica*, F. McCoy, A. M. N. H. ser. 2, VIII
(1851), p. 400, and Contrib. Brit. Pal. (1854), p. 222, and Brit.
Palæoz. Foss. (1852), pl. i H, f. 29, 29 *a, b*, p. 232. *Middle
Bala; Craig Head, Ayrshire.*

O. calligramma, Dalman, var. **calliptycha**, F. McCoy, Brit.
Palæoz. Foss. (1852), p. 215. *Middle Bala; Llansaintffraid.*

O. crispa, F. McCoy, Brit. Palæoz. Foss. (1855), pl. i H, f. 43,
43 *a, b*, p. 216. *Middle Bala; Bala.*

O. filosa, v. Strophomena filosa.

O. flabellulum, Sowerby, T. Davidson, Brit. Foss. Brach. III,
part vii (1869), pl. xxxiv, f. 12 *a*, p. 248. *Bala Beds; Boduan,
Carnarvonshire.*

O.? **hirnantensis**, F. McCoy, A. M. N. H. ser. 2, VIII (1851),
p. 395, and Contrib. Brit. Pal. (1854), p. 217, and Brit. Palæoz.
Foss. (1852), pl. i H, f. 11, 11 *a—c*, p. 219. T. Davidson, Brit.
Foss. Brach. III, part vii (1869), pl. xxxii, f. 5—7, p. 261.
Hirnant Limestone; Maes Hir and Aber Hirnant.

O. interlineata, J. de C. Sowerby, Trans. Geol. Soc. ser. 2, v
(1840), pl. liii, f. 11, pl. liv, f. 14. *Atrypa striatula*, J. de
C. Sowerby, *ibid.* (1840), pl. liv, f. 10. *Upper Devonian;
Petherwin.*

Orthis michelini ? (L'Eveillé), T. Davidson, Brit. Foss. Brach. ii, part v (1861), pl. xxx, f. 12, p. 132. *Carboniferous Limestone; Settle.* Burrow Collection.

O. pecten, v. Strophomena pecten.

O. persarmentosa, v. Streptorhynchus persarmentosus.

O. porcata, F. McCoy, Brit. Palæoz. Foss. (1852), pl. i H, f. 41 *a—c*, 42, 42 *a*, p. 223. T. Davidson, Brit. Foss. Brach. iii, part vii (1869), pl. xxxi, f. 15, p. 250. *Coniston Limestone; Coniston.*

O. retrorsistria, F. McCoy, A. M. N. H. ser. 2, viii (1851), p. 396, and Contrib. Brit. Pal. (1854), p. 218, and Brit. Palæoz. Foss. (1852), pl. i H, f. 12—14, p. 224. T. Davidson, Brit. Foss. Brach. v (1883), pl. xiv, f. 7—16 (f. 8 and 14 missing), p. 185. *O. alternata*, Salter (*non* Sowerby), Cat. (1873), p. 60. *O. alternata* var. *retrorsistria*, T. Davidson, Brit. Foss. Brach. iii, part vii (1869), pl. xxxvi, f. 39—42, p. 265. *Middle Bala; Cerrig-y-Druidion, Hafod Evan, Pen-y-Gaer, Pentre Cwmdu.*

O. sagittifera, F. McCoy, A. M. N. H. ser. 2, viii (1851), p. 398, and Contrib. Brit. Pal. (1854), p. 220, and Brit. Palæoz. Foss. (1852), pl. i H, f. 15—19, p. 227. T. Davidson, Brit. Foss. Brach. iii, part vii (1869), pl. xxxvi, f. 18—21, p. 260. *Bala Beds; Aber Hirnant.*

O. sarmentosa, F. McCoy, Brit. Palæoz. Foss. (1852), pl. i H, f. 25—28, p. 227. T. Davidson, Brit. Foss. Brach. iii, part vii (1869), pl. xxxvi, f. 35—38, p. 262, and *ibid.* v (1883), p. 189. *Orthis testudinaria*, Dalman, Salter, Cat. (1873), p. 60. *Middle Bala; Llyn Ogwen.*

O.? spiriferoides (McCoy). *Strophomena spiriferoides*, F. McCoy, A. M. N. H. ser. 2, viii (1851), p. 402, and Contrib. Brit. Pal. (1854), p. 224, and [*Leptœna*] Brit. Palæoz. Foss. (1852), p. 246, woodcuts. *Bala Beds; Moel-y-Garth, Welshpool, Horderley, Allt-yr-Anker, etc.*

O. turgida, F. McCoy, A. M. N. H. ser. 2, viii (1851), p. 399, and Contrib. Brit. Pal. (1854), p. 221, and Brit. Palæoz. Foss. (1852), pl. i H, f. 20, 20 *a* (Llandeilo), 21 (Golden Grove), 22 (Craig-y-Beri), 23, 24 (Conway Falls), p. 229. T. Davidson, Brit. Foss. Brach. v (1883), pl. xiv, f. 17, 19, 19 *a*, 20, 20 *a*,

p. 187. *Llandeilo Beds; Golden Grove, and Llandeilo. Middle Bala; Conway Falls.*

Orthisina adscendens (Pander), T. Davidson, Brit. Foss. Brach. III, part vii (1871), pl. xlix, f. 27, p. 278. *Bala Beds; Cefn-coedog, Corwen.*

Orthisina scotica, v. Orthis calligramma *var.* scotica.

Pentamerus carbonarius, v. Cyrtina carbonaria.

Pentamerus knighti, Sowerby, var. **elongatus**, F. McCoy, Brit. Palæoz. Foss. (1855), p. 209. *Aymestry Limestone; Woolhope.*

P. microcamerus, v. Stricklandinia lens.

Porambonites? intercedens, Pander, T. Davidson, Brit. Foss. Brach. III, part vii (1869), pl. xxvi, f. 3, p. 195 (not pl. xxv, f. 17—19, v. Rhynchonella maccoyana). *Bala Beds; Wrae Quarry, Peebleshire.* This specimen cannot be found.

Productus elegans, v. P. punctatus *var.* elegans.

Productus prælongus (Sowerby), F. McCoy, Brit. Palæoz. Foss. (1852), pl. ii A, f. 10, p. 390. T. Davidson, Brit. Foss. Brach. III, part vii (1865), pl. xix, f. 22, p. 102. *Leptæna prælonga*, J. de C. Sowerby, Trans. Geol. Soc. ser. 2, v (1840), pl. liii, f. 29. *Upper Devonian; Croyde Bay.*

P. proboscideus, De Verneuil, T. Davidson, Brit. Foss. Brach. II, part v (1861), pl. xxxiii, f. 2, p. 163. *Carboniferous Limestone; Settle.* Burrow Collection. The specimen does not agree well with the figure.

P. punctatus (Martin) var. **elegans**, McCoy. *P. elegans*, F. McCoy, Brit. Palæoz. Foss. (1855), pl. iii H, f. 4, 4 *a—c*, p. 460. *Carboniferous Limestone; Lowick.*

P. subaculeatus, Murchison. *Leptæana fragaria* var., J. de C. Sowerby, Trans. Geol. Soc. ser. 2, v (1840), pl. liv, f. 3. *Upper Devonian; Petherwin.*

Pseudocrania divaricata, v. Crania divaricata.

Retzia radialis (Phillips), T. Davidson, Brit. Foss. Brach. II, part v (1861), pl. xvii, f. 20 *a—d*, 21, p. 87. *Carboniferous Limestone; Settle.* Burrow Collection.

Rhynchonella acuminata (Martin), T. Davidson, Brit. Foss.
Brach. II, part V (1861), pl. xx, f. 5, 7, 8, p. 93. *Carboniferous
Limestone; Settle.* Burrow Collection.

R. acuminata (Martin) var. **mesogona** (Phillips). *Atrypa triangularis,* J. de C. Sowerby, Trans. Geol. Soc. ser. 2, V (1840),
pl. liv, f. 9. *Rhynchonella triangularis* (Sowerby), T. Davidson,
Brit. Foss. Brach. III, part VI (1865), pl. xiii, f. 5, p. 60. *Hemithyris acuminata* var. *mesogona* (Phillips), F. McCoy, Brit.
Palæoz. Foss. (1852), p. 381. *Upper Devonian; Petherwin.*

R. angulata (Linnæus), T. Davidson, Brit. Foss. Brach. II, part
V (1861), pl. xix, f. 16, 16 *a, b,* p. 107. *Carboniferous Limestone; Settle, Yorkshire.* Burrow Collection.

R. antidichotoma (Buvignier), T. Davidson, Brit. Foss. Brach.
IV (1874), pl. viii, f. 20, p. 65. *Lower Greensand; Upware.*
The specimen does not quite agree with the figure.

R. cuboides (Sowerby). *Atrypa cuboides,* J. de C. Sowerby,
Trans. Geol. Soc. ser. 2, V (1840), pl. lvi, f. 24. *Devonian;
Plymouth.*

R. elliptica?, Schnur, T. Davidson, Brit. Foss. Brach. IV (1881),
pl. xxxviii, f. 23, p. 342. *Budleigh Salterton Pebble Bed.*
Presented by A. Wyatt Edgell, Esq.

R. flexistria (Phillips), T. Davidson, Brit. Foss. Brach. II, part
V (1861), pl. xxiv, f. 5, 5 *a,* p. 105. *Hemithyris heteroptycha,*
F. McCoy, A. M. N. H. ser. 2, X (1852), p. 424, and Contrib.
Brit. Pal. (1854), p. 251, and Brit. Palæoz. Foss. (1855), pl. iii D,
f. 19 *a—c,* p. 440. *Carboniferous Limestone; Derbyshire.*

R. hopkinsi [McCoy MS.], T. Davidson, Brit. Foss. Brach. I, part
iii (1852), pl. A, f. 20, 20 *a,* 21, 21 *a, d,* p. 97, and Appendix
p. 21. *Great Oolite; Minchinhampton?* (Probably from
France.)

R. leedsi [Walker MS.], T. Davidson, Brit. Foss. Brach. IV (1878),
pl. xxviii, f. 17, 17 *a, b,* p. 216. *Cornbrash; Scarborough.*
Leckenby Collection.

R.? maccoyana, T. Davidson, Brit. Foss. Brach. V (1883), p. 161.
Porambonites intercedens, Pander, T. Davidson, *ibid.* III, part
vii (1869), pl. xxv, f. 17—19 (not pl. xxvi, f. 3), p. 195. *Bala*

Beds; Wrae Quarry, Peebleshire. Fig. 17 is the only specimen which can be identified with certainty.

Rhynchonella multiformis (Roemer), T. Davidson, Brit. Foss. Brach. IV (1874), pl. viii, f. [22 ? missing] 23, p. 63. *Lower Greensand (Tealby Series); Claxby.*

R. nasuta (McCoy), T. Davidson, Brit. Foss. Brach. III, part vii (1869), pl. xxiii, f. 19, 19 *a, b,* p. 173. *Hemithyris nasuta,* F. McCoy, A. M. N. H. ser. 2, VIII (1851), p. 393, and Contrib. Brit. Pal. (1854), p. 215, and Brit. Palæoz. Foss. (1852), pl. i L, f. 5, 5 *a,* p. 203. *Middle Bala; Craig Head, Girvan.*

R. pleurodon (Phillips). *Atrypa fallax,* J. de C. Sowerby, Trans. Geol. Soc. ser. 2, v (1840), pl. liv, f. 15. *Upper Devonian; Petherwin.*

R. subdentata (Sowerby), T. Davidson, Brit. Foss. Brach. III, part vi (1865), pl. xiii, f. 15, p. 62. *Atrypa subdentata,* J. de C. Sowerby, Trans. Geol. Soc. ser. 2, v (1840), pl. liv, f. 7. *Hemithyris subdentata,* F. McCoy, Brit. Palæoz. Foss. (1852), p. 383. *Devonian; Petherwin.*

R. subtetrahedra, T. Davidson, Brit. Foss. Brach. I, part iii (1852), pl. xvi, f. 10, p. 95. *Inferior Oolite; Dundry.* Walton Collection. This specimen is missing.

R. speetonensis, T. Davidson, Brit. Foss. Brach. IV (1874), pl. viii, f. 32, p. 69. *Speeton Clay; Speeton.* This specimen is missing.

R. tetrahedra, v. Rhynchonella subdecorata.

R. subdecorata, T. Davidson, Brit. Foss. Brach. I, part iii, Appendix (1852), p. 21. *R. tetrahedra* (Sowerby), T. Davidson, Brit. Foss. Brach. I, part iii (1852), pl. xviii, f. 10, p. 93. *Inferior Oolite; Cheltenham.* Walton Collection. This specimen is missing.

R. walkeri, T. Davidson, Brit. Foss. Brach. IV (1874), pl. viii, f. 33, p. 68. *Lower Greensand (Tealby Series); Claxby.*

R. wilsoni (Sowerby) var. **davidsoni** (McCoy). *Hemithyris davidsoni,* F. McCoy, A. M. N. H. ser. 2, VIII (1851), p. 392, and Contrib. Brit. Pal. (1854), p. 214, and Brit. Palæoz. Foss. (1852), p. 200. *Upper Ludlow; Burton and Brockton.*

Rhynchonella wilsoni (Sowerby) var. **sphæroidalis** (McCoy), T. Davidson, Brit. Foss. Brach. III, part vii (1869), pl. xxiii, f. 10, 10 a, b, p. 173. *Hemithyris sphæroidalis*, F. McCoy, A. M. N. H. ser. 2, VIII (1851), p. 393, and Contrib. Brit. Pal. (1854), p. 215, and Brit. Palæoz. Foss. (1855), pl. i L, f. 4, 4 a, b, p. 206. *Aymestry Limestone; Botvylle, near Church Stretton.*

R. sp., Salter, Cat. (1873), p. 140. *Wenlock Limestone; Dudley.* Fletcher Collection.

R. sp., T. Davidson, Brit. Foss. Brach. IV (1874), pl. viii, f. 26. *Greensand; Blackdown.*

Seminula ficus, v. Terebratula hastata *var.* ficus.

S. juvenis, v. Terebratula sp.

S. virgoides, v. Terebratula hastata.

Siphonotreta micula, F. McCoy, A. M. N. H. ser. 2, VIII (1851), p. 389, and Contrib. Brit. Pal. (1854), p. 211, and Brit. Palæoz. Foss. (1852), pl. i H, f. 3, 3 a, p. 188. T. Davidson, Brit. Foss. Brach. III, part vii (1866), pl. viii, f. 2, 2 a, p. 76. *Llandeilo Beds; Pen-cerrig, near Builth.*

Spirifera costata, v. Spirifera speciosa.

S. disjuncta, v. Spirifera verneuili.

Spirifera duplicostata, Phillips, T. Davidson, Brit. Foss. Brach. II, part v (1858), pl. iv, f. 11 a, b, c, p. 24. *Spirifera fasciculata*, F. McCoy, A. M. N. H. ser. 2, x (1852), p. 422, and Contrib. Brit. Pal. (1854), p. 249, and Brit. Palæoz. Foss. (1855), pl. iii D, f. 25 a, b, c, p. 416. *Carboniferous Limestone; Derbyshire.*

S. fasciculata, v. Spirifera duplicostata.

S. gigantea, v. Spirifera verneuili.

S. grandicostata, v. Spirifera trigonalis.

S. integricostata, Phillips (? var.), T. Davidson, Brit. Foss. Brach. II, part v (1858), pl. iv, f. 12 a, b, p. 55. *S. paucicostata*, F. McCoy, A. M. N. H. ser. 2, x (1852), p. 423, and Contrib. Brit. Pal. (1854), p. 250, and Brit. Palæoz. Foss. (1855), pl. iii D, f. 26 a, b, p. 420. *Carboniferous Limestone; Derbyshire.*

S. laminosa, v. Spiriferina laminosa.

Spirifera mosquensis, v. Spirifera striata.

S. ornithorhyncha, v. Spirifera triangularis.

S. ovalis, Phillips, T. Davidson, Brit. Foss. Brach. II, part V (1859), pl. ix, f. 21, p. 53. *Carboniferous Limestone; Derbyshire.* Presented by W. Hopkins, Esq.

S. ovalis, Phillips, var. **hemispherica**, F. McCoy, Brit. Palæoz. Foss. (1855), pl. iii D, f. 28, 28 *a*, *b*, p. 419. *Carboniferous Limestone; Isle of Man.*

S. paucicostata, v. Spirifera integricostata.

S. plicatella (Linnæus) var. **globosa**, Salter, T. Davidson, Brit. Foss. Brach. III, part vii (1866), pl. ix, f. 7, 7 *a*, *b*, p. 89. *Wenlock Limestone; Dudley.* Fletcher Collection. This specimen is missing.

S. plicatella (Linnæus) var. **radiata**, Sowerby, T. Davidson, Brit. Foss. Brach. III, part vii (1866), pl. ix, f. 2, 2 *a*, *b*, p. 87. *Wenlock Limestone; Dudley.* Fletcher Collection. This specimen is missing.

S. speciosa (Schlotheim). *S. costata*, J. de C. Sowerby, Trans. Geol. Soc. ser. 2, v (1840), pl. lv, f. 5—7. *S. speciosa?* T. Davidson, Brit. Foss. Brach. III, part vi (1864), pl. viii, f. 9, p. 29. *Devonian; Fowey.*

S. striata (Martin), T. Davidson, Brit. Foss. Brach. II, part v (1858), pl. ii, f. 20, p. 19. *Carboniferous Limestone; Ardnaglass.* Another specimen, T. Davidson, *ibid.* II, part v (1863), pl. lii, f. 1, p. 221. *Carboniferous Limestone; Settle, Yorkshire.* Burrow Collection.

S. mosquensis (Fischer), T. Davidson, Brit. Foss. Brach. II, part v (1858), pl. iv, f. 13, p. 22. *Carboniferous Limestone; Derbyshire.* This specimen is missing.

S. triangularis (Martin), T. Davidson, Brit. Foss. Brach. II, part v (1863), pl. l, f. 10—18, p. 223. *Carboniferous (Lower Scar) Limestone; Settle.* Burrow Collection. Figs. 10, 13, and 18 cannot be found.

S. ornithorhyncha, F. McCoy, Brit. Palæoz. Foss. (1855), pl. iii D, f. 27, 27 *a*, p. 418. *Carboniferous Limestone; Derbyshire.*

Spirifera trigonalis (Martin), T. Davidson, Brit. Foss. Brach.
II, part v (1863), pl. l, f. 6, 7, p. 222. *Carboniferous Limestone;*
Settle, Yorkshire. Burrow Collection.

S. grandicostata, F. McCoy, A. M. N. H. ser. 2, x (1852),
p. 422, and Contrib. Brit. Pal. (1854), p. 250, and Brit. Palæoz.
Foss. (1855), pl. iii D, f. 29 *a, b,* p. 417. T. Davidson, Brit.
Foss. Brach. II, part v (1858), pl. v, f. 38, 39, p. 33. Probably
a variety of S. trigonalis (Martin), Davidson, *ibid.* p. 222, and
ibid. IV (1880), p. 276. *Carboniferous Limestone; Derbyshire.*

S. urei, Fleming, T. Davidson, Brit. Foss. Brach. III, part vi
(1864), pl. iv, f. 26, p. 41. *Atrypa unguiculus*, J. de C. Sowerby,
Trans. Geol. Soc. ser. 2, v (1840), pl. liv, f. 8. *Upper Devonian;*
Petherwin.

S. verneuili, Murchison. *S. disjuncta*, J. de C. Sowerby, Trans.
Geol. Soc. ser. 2, v (1840), pl. liii, f. 8 (Barnstaple), pl. liv,
f. 12, 13 (Petherwin). T. Davidson, Brit. Foss. Brach. III,
part vi (1864), pl. v, f. 1 (Petherwin), f. 2 (Barnstaple), p. 23.
S. gigantea, J. de C. Sowerby, *loc. cit.* pl. lv, f. 1—4 (Tintagel).
Devonian; Barnstaple, Petherwin, and Tintagel.

Spiriferina cristata (Schlotheim) var. **biplicata**, T. Davidson,
Brit. Foss. Brach. II, part v (1863), pl. lii, f. 11, [12? missing],
p. 226. *Carboniferous Limestone; Settle.* Burrow Collection.

S. laminosa (McCoy). *Spirifer laminosa*, F. McCoy, Brit. Palæoz.
Foss. (1855), p. 426. T. Davidson, Brit. Foss. Brach. II, part
v, (1858), pl. vii, f. 21, 22, p. 36. *Carboniferous Limestone;*
Derbyshire.

Spondylobolus craniolaris, v. Athyris obovata.

Streptorhynchus gigas (McCoy), T. Davidson, Brit. Foss.
Brach. III, part vi (1865), pl. xvi, f. 1, p. 83. *Strophomena*
gigas, F. McCoy, A. M. N. H. ser. 2, VIII (1851), p. 485, and
Contrib. Brit. Pal. (1854), p. 236. *Leptæna (Strophomena)*
gigas, F. McCoy, Brit. Palæoz. Foss. (1852), pl. ii A, f. 7, p. 386.
Devonian; Looe.

Streptorhynchus? **persarmentosus** (McCoy), T. Davidson,
Brit. Foss. Brach. III, part vi (1865), pl. xvi, f. 5, p. 84. *Orthis*
persamentosa, F. McCoy, A. M. N. H. ser. 2, VIII (1851), p. 484,

and Contrib. Brit. Pal. (1854), p. 235, and Brit. Palæoz. Foss. (1852), pl. ii A, f. 9, p. 385. *Devonian; Polruan, Cornwall.*

Stricklandinia lens (Sowerby), T. Davidson, Brit. Foss. Brach. III, part vii (1867), pl. xix, f. 23, p. 161. *Pentamerus microcamerus,* F. McCoy, A. M. N. H. ser. 2, VIII (1851), p. 290, and Contrib. Brit. Pal. (1854), p. 212, and Brit. Palæoz. Foss. (1852), p. 210, woodcuts. *Llandovery Beds; Mandinam, and May Hill.*

Stringocephalus burtini, Defrance, T. Davidson, Brit. Foss. Brach. III, part vi (1864), pl. ii, f. 9, 9 *a*, p. 11. *Uncites lævis,* F. McCoy, A. M. N. H. ser. 2, VIII (1851), p. 483, and Contrib. Brit. Pal. (1854), p. 234, and Brit. Palæoz. Foss. (1852), pl. ii A, f. 6 *a—c*, p. 380. *Devonian; Newton Bushel.*

Another specimen, T. Davidson, *ibid.* (1864), pl. ii, f. 6, 6 *a, b*, p. 11. *Devonian; Woolborough Quarry, Newton Abbot.* Walton Collection. This specimen is missing.

Strophomena deltoidea (Conrad) var. **undata** (McCoy), T. Davidson, Brit. Foss. Brach. III, part vii (1871), pl. xxxix, f. 23, 24, p. 295. *Leptæna deltoidea* var. *undata,* F. McCoy, Brit. Palæoz. Foss. (1852), pl. i H, f. 38, 39, 39 *a, b, c*, p. 234. *Middle Bala; fig.* 38 *from Llwyn-y-Ci, f.* 39, 39 *a, b, from Llandeilo, f.* 39 *c from Bryn-melyn, near Bala.*

S. filosa (Sowerby). *Orthis filosa,* T. Davidson, London Geol. Journal, part ii (1847), pl. xiii, f. 24, p. 62. *Wenlock Limestone; Dudley.* Fletcher Collection. The specimen cannot be found.

S. gigas, v. Streptorhynchus gigas.

S. nobilis, v. Leptæna? nobilis.

S. pecten (Linnæus). *Orthis pecten,* T. Davidson, London Geol. Journal, part ii (1847), pl. xiii, f. 18—23, p. 61. *Wenlock Limestone; Dudley.* Fletcher Collection. The specimens cannot be found.

S. simulans, F. McCoy, A. M. N. H. ser. 2, VIII (1851), p. 403, and Contrib. Brit. Pal. (1854), p. 225, and [*Leptæna*] Brit. Palæoz. Foss. (1852), pl. i H, f. 33, 34, 34 *a*, [f. 35 missing] p. 246. T. Davidson, Brit. Foss. Brach. III, part vii (1871), pl. xlii, f. 9, 9 *a—c*, 10, p. 297. *Middle Bala; Cefn Coch. Wenlock Beds; Myddelton Park, Caermarthen.*

Strophomena spiriferoides, v. Orthis ? spiriferoides.

S. ungula (McCoy). *Leptæna* (*Leptagonia*) *ungula*, F. McCoy, A. M. N. H. ser. 2, VIII (1851), p. 404, and Contrib. Brit. Pal. (1854), p. 226, and Brit. Palæoz. Foss. (1852), pl. i H, f. 36, 36 *a*, *b*, 37, 37 *a*, p. 249. *Middle Bala; Llansaintffraid, Glyn Ceiriog.*

Terebratella keepingi [Walker MS.], W. Keeping, Foss. Upware and Brickhill (1883), pl. vii, f. 19 *a—d*, p. 130. T. Davidson, Brit. Foss. Brach. V (1884), pl. xviii, f. 4, 4 *a*, 5, p. 245. *Lower Greensand; Brickhill.*

T. menardi (Lamarck), T. Davidson, Brit. Foss. Brach. V (1884), pl. xviii, f. 6, p. 247. *Lower Greensand; Upware.* Presented by E. Towry Whyte, Esq., M.A.

Terebratula boloniensis, Sauvage and Rigaux, T. Davidson, Brit. Foss. Brach. IV (1878), pl. xix, f. 2, 2 *a*, p. 154. *Coral Rag; Malton.* Leckenby Collection.

T. capillata, d'Archiac, T. Davidson, Brit. Foss. Brach. IV (1874), pl. vii, f. 2, 2 *a—c*, p. 33. *Lower Greensand; Upware.*

T. dallasi, Walker, T. Davidson, Brit. Foss. Brach. IV (1874), pl. iii, f. 1, 1 *a*, 2, 2 *a*, *b*, p. 45. *Lower Greensand; Upware.*

T. depressa, Lamarck, T. Davidson, Brit. Foss. Brach. IV (1874), pl. iv, f. 3, p. 40. *Lower Greensand; Acre House, Lincolnshire.* This specimen is missing.

Another specimen, T. Davidson, Brit. Foss. Brach. V (1884), pl. xviii, f. 7, p. 251. *Lower Greensand; Brickhill.*

T. edwardsi, v. T. punctata *var.* edwardsi.

T. galeiformis [McCoy MS.], T. Davidson, Brit. Foss. Brach. I, Appendix (1855), pl. A, f. 15, 15 *a*, *b*, p. 19. *Inferior Oolite; Brimscombe, near Minchinhampton.*

T. globata, Sowerby, T. Davidson, Brit. Foss. Brach. I, part iii (1851), pl. xiii, f. 5, p. 54. *Inferior Oolite; Cheltenham.* Walton Collection.

T. (Dielasma) hastata, Sowerby, T. Davidson, Brit. Foss. Brach. II, part v (1858), pl. i, f. 12, 12 *a—d*, p. 11. *Seminula virgoides*,

F. McCoy, Brit. Palæoz. Foss. (1855), pl. iii D, f. 23 *a—d,* p. 413. *Carboniferous Limestone; Derbyshire.*

Terebratula (Dielasma) hastata, Sowerby, var. **ficus** (McCoy), T. Davidson, Brit. Foss. Brach. II, part V (1858), pl. i, f. 15, *a, b,* p. 13. *Seminula ficus,* F. McCoy, A. M. N. H. ser. 2, X (1852), p. 421, and Contrib. Brit. Pal. (1854), p. 248, and Brit. Palæoz. Foss. (1855), pl. iii D, f. 22, *a, b,* p. 409. *Carboniferous Limestone; Derbyshire.*

T. indentata, v. Waldheimia indentata.

T. insignis, Schübler, var. **maltonensis,** Oppel, T. Davidson, Brit. Foss. Brach. IV (1876), pl. xv, f. 5, 5 *a,* p. 126. *Coral Rag; Malton.* Leckenby Collection.

T. lankesteri, Walker, T. Davidson, Brit. Foss. Brach. IV (1874), pl. iii, f. 10, 10 *a, b,* p. 38. *Lower Greensand; Upware.*

T. moutoniana, d'Orbigny, var. **brickhillensis,** W. Keeping, Foss. Upware and Brickhill (1883), p. 162. T. Davidson, Brit. Foss. Brach. V (1884), pl. xviii, f. 8, p. 251. *Lower Greensand; Brickhill.*

T. ovata ?, Sowerby, T. Davidson, Brit. Foss. Brach. IV (1874), pl. vii, f. 1, 1 *a, b, c,* p. 32. *Greensand; Blackdown.*

T. ovoides, Sowerby, var. **rex,** Ray Lankester, W. Keeping, Foss. Upware and Brickhill, pl. viii, f. 10 *a, b,* p. 154. *Lower Greensand (derived); Upware.*

T. punctata, Sowerby, var. **edwardsi,** T. Davidson, Brit. Foss. Brach. IV (1876), p. 131. *T. edwardsi,* T. Davidson, *ibid.* I, part iii (1851), pl. vi, f. 11, 11 *a,* p. 30. *Middle Lias; near Ilminster.* Walton Collection.

T. sella, Sowerby, **var.,** T. Davidson, Brit. Foss. Brach. IV (1874), pl. v, f. 11, 11 *a,* p. 36. *Lower Greensand (Tealby Series); Claxby.*

T. simplex, Buckman, T. Davidson, Brit. Foss. Brach. I, part iii (1851), pl. viii, f. 3, 3 *a,* p. 48. *Inferior Oolite; Leckhampton.* Walton Collection.

T. tawneyi, G. F. Whidborne, Q. J. G. S. XXXIX (1883), pl. xix, f. 12, 12 *a, b,* p. 536. T. Davidson, Brit. Foss. Brach. V (1884),

pl. xviii, f. 14, 14 *a*, *b*, p. 255. *Inferior Oolite; Dundry.*
Presented by the Rev. G. F. Whidborne, M.A.

Terebratula (Dielasma) vescicularis, De Koninck, T. David-
son, Brit. Foss. Brach. II, part v (1858), pl. ii, f. 7, p. 15.
Seminula seminula, F. McCoy (*non* Phillips), Brit. Palæoz.
Foss. (1855), p. 412. *Carboniferous Limestone; Derbyshire.*

T. sp.? T. Davidson, Brit. Foss. Brach. II, part v (1858), pl. i, f. 17,
17 *a.* *Seminula juvenis* (Sowerby), F. McCoy, Brit. Palæoz.
Foss. (1855), p. 410. *Carboniferous Limestone; Derbyshire.*

Triplesia? **maccoyana**, T. Davidson, Brit. Foss. Brach. III,
part vii (1869), pl. xxiv, f. 29, p. 199. *Bala Limestone;
Bryn-bedwog Quarry, near Bala.*

Uncites lœvis, v. Stringocephalus burtini.

Waldheimia (Zeilleria) bonneyi, W. Keeping, Foss. Upware
and Brickhill (1883), pl. vii, f. 4 *a—c*, p. 129. T. Davidson,
Brit. Foss. Brach. v (1884), pl. xviii, f. 9, 9 *a*, *b*, p. 247. *Lower
Greensand; Brickhill.*

W. (Zeilleria) bucculenta (Sowerby), T. Davidson, Brit. Foss.
Brach. IV (1878), pl. xxii, f. 23. *Lower Calcareous Grit; near
Malton (probably Appleton).* Leckenby Collection.
Another specimen, T. Davidson, Brit. Foss. Brach. IV (1878),
pl. xxiv, f. 23, 23 *a*, p. 173. *Elsworth Rock; Elsworth.*

W. (Zeilleria) faba (d'Orbigny), T. Davidson, Brit. Foss. Brach.
IV (1874), pl. vi, f. 14, p. 55. *Lower Greensand (Tealby
Series); Claxby.*

W. (Zeilleria) indentata (Sowerby). *Terebratula indentata,* T.
Davidson, Brit. Foss. Brach. I, part iii (1851), pl. v, f. 25, 25 *a*,
b, 26, p. 46. *Lias; Farrington Gurney, Somerset.* Walton
Collection. The specimens do not agree well with the figures.

W. (Zeilleria) juddi, Walker, T. Davidson, Brit. Foss. Brach. IV
(1874), pl. vii, f. 15, p. 50. *Lower Greensand; Upware.* This
specimen is missing.

W. (Zeilleria) obovata (Sowerby) var. **stiltonensis** [Walker
MS.], T. Davidson, Brit. Foss. Brach. IV (1878), pl. xxii, f. 20,
p. 172. *Cornbrash; Scarborough.* Leckenby Collection.
Waldheimia bucculenta? (Sowerby), T. Davidson, *ibid.* IV

(1878), pl. xxiv, f. 25, 25 *a*, p. 173. *Cornbrash; Scarborough.* Leckenby Collection.

Waldheimia (Zeilleria) tamarindus (Sowerby) var. **tilbyensis**, T. Davidson, Brit. Foss. Brach. IV (1874), pl. vii, f. 5, p. 49. *Lower Greensand (Tealby Series); Claxby.*

W. (Aulacothyris) walkeri, T. Davidson, Brit. Foss. Brach. IV (1874), pl. vi, f. 8, p. 54. *Lower Greensand (Tealby Series); Claxby.*

W. (Zeilleria) wanklyni, Walker, var. **angusta**, Walker, T. Davidson, Brit. Foss. Brach. IV (1874), pl. vii, f. 26, p. 51. *Lower Greensand; Upware.* This specimen is missing.

MOLLUSCA.

LAMELLIBRANCHIATA.

Ambonychia? **acuticostata**, F. McCoy, A. M. N. H. ser. 2, VII (1851), p. 59, and Contrib. Brit. Pal. (1854), p. 194, and Brit. Palæoz. Foss. (1852), pl. i K, f. 16, 16 *a*, p. 264. *Lower Ludlow; Dinas Bran, Llangollen.*

Anatina siliqua (Agassiz), J. Lycett, Supp. Mon. Moll. Gt. Ool. etc. (1863), pl. xxxv, f. 15, p. 83. *Cornbrash; Scarborough.* Leckenby Collection.

A. versicostata, Buvignier, J. Leckenby, Q. J. G. S. xv (1859), pl. iii, f. 6. *Kellaways Rock; Scarborough.* Leckenby Collection.

Anodontopsis angustifrons, F. McCoy, A. M. N. H. ser. 2, VII (1851), p. 54, and Contrib. Brit. Pal. (1854), p. 189, and Brit. Palæoz. Foss. (1852), pl. i K, f. 14, 15, p. 271. *Upper Ludlow; Benson Knot, Kendal.*

A. bulla (McCoy), F. McCoy, Brit. Palæoz. Foss. (1852), pl. i K, f. 11, 12, p. 271. *Upper Ludlow; Kirkby Moor, Kendal.* The specimen does not agree very well with the figure.

Anodontopsis quadrata, F. McCoy, A. M. N. H. ser. 2, VII (1851), p. 55, and Contrib. Brit. Pal. (1854), p. 190, and Brit. Palæoz. Foss. (1852), pl. i K, f. 10, p. 272. *Palœarca diagona*, Salter, Cat. (1873), p. 190. *Upper Ludlow Tilestones; Storm Hill, Llandeilo.*

A. securiformis, F. McCoy, A. M. N. H. ser. 2, VII (1851), p. 55, and Contrib. Brit. Pal. (1854), p. 190, and Brit. Palæoz. Foss. (1852), pl. i L, f. 9, 9 *a*, p. 272. *Pseudaxinus securiformis*, Salter, Cat. (1873), p. 183. *Upper Ludlow; Benson Knot, Kendal.*

Anomia papyracea, d'Orbigny, var. **burwellensis**, R. Etheridge, Appendix to W. H. Penning and A. J. Jukes-Browne, Geol. Cambridge (1881), pl. iii. f. 3, 4, p. 145. *Lower Chalk (Totternhoe Stone); Burwell.*

Arca carteroni, d'Orbigny, W. Keeping, Foss. Upware and Brickhill (1883), pl. v, f. 7 *a, b*, p. 114. *Lower Greensand; Upware.*

A. edmondiiformis, F. McCoy, A. M. N. H. ser. 2, VII (1851), p. 52, and Contrib. Brit. Pal. (1854), p. 187, and Brit. Palæoz. Foss. (1852), pl. i K, f. 2, 3, p. 283. *Fig. 2 from Denbigh Flags, near Llangynyw; fig. 3 from the Middle Bala, Allt-yr-Gader.*

A. laekeniana, Le Hon, var. **cylindrica**, S. V. Wood, Eoc. Bivalves I (1864), pl. xv, f. 7 *c*, p. 80. *Bracklesham Beds; Huntingbridge.* Fisher Collection. This specimen is missing.

A. (Byssoarca) subæqualis, F. McCoy, A. M. N. H. ser. 2, VII (1851), p. 51, and Contrib. Brit. Pal. (1854), p. 186, and Brit. Palæoz. Foss. (1852), pl. i K, f. 1, 1 *a*, p. 283. *Ctenodonta subæqualis*, Salter, Cat. (1873), p. 190. *Upper Ludlow; Llechclawdd, Myddfai, near Llandovery.*

A. ? sulcata v. Inoceramus sulcatus.

Astarte aytonensis [Bean MS.], J. Lycett, Supp. Mon. Moll. Great Ool. etc. (1863), pl. xl, f. 13, p. 78. *Great Oolite; Combe Down, near Bath.* Walton Collection.

Another specimen, J. F. Blake and W. H. Hudleston, Q. J. G. S. XXXIII (1877), pl. xiv, f. 11, p. 397. *Coralline Oolite; Ayton, near Scarborough.* Leckenby Collection.

A. bathonica, J. Lycett, Supp. Mon. Moll. Gt. Ool. etc. (1863), pl. xl, f. 23, 23 *a*, p. 76. *Great Oolite; Hampton Cliffs, near Bath.* Walton Collection.

Astarte crebricostata ?, Forbes, S. V. Wood, Crag Mollusca, part ii (1850), pl. xvii, f. 6, p. 186. *Bridlington Crag; Bridlington.* Leckenby Collection.

A. elegans, Sowerby, J. Morris and J. Lycett, Mollusca Gt. Ool. part ii (1853), pl. xiv, f. 14, p. 86. *Millepore Bed; Stainton Dale Cliffs.* Leckenby Collection.

A. elegans, Sowerby, var. **munda**, G. F. Whidborne, Q. J. G. S. xxxix (1883), pl. xix, f. 4, p. 527. *Inferior Oolite; Hardington, Somerset.* Presented by the Rev. G. F. Whidborne, M.A.

A. fimbriata [Walton MS.], J. Lycett, Supp. Mon. Moll. Great Ool. etc. (1863), pl. xl, f. 34, 34 *a*, p. 77. *Forest Marble; Farleigh.* Walton Collection.

A. hilpertensis, J. Lycett, Supp. Mon. Moll. Gt. Ool. etc. (1863), pl. xxxvi, f. 10, p. 78. *Cornbrash; Hilperton, Wilts.* Walton Collection. This specimen is missing.

A.? **ignota**, J. Lycett, Supp. Mon. Moll. Gt. Ool. etc. (1863), pl. xl, f. 10, p. 77. *Forest Marble; Laycock, Wilts.* Walton Collection.

A. mutabilis, S. V. Wood, Crag Mollusca, part ii (1850), pl. xvi, f. 1 *e*, *f*, p. 179. *Bridlington Crag; Bridlington.* Leckenby Collection.

A. politula, Bean, J. Lycett, Supp. Mon. Moll. Gt. Ool. etc. (1863), pl. xxxv, f. 16, p. 73. *Cornbrash; Scarborough.* Leckenby Collection.

A. robusta, J. Lycett, Supp. Mon. Moll. Gt. Ool. etc. (1863), pl. xxxv, f. 6, 6 *a*, p. 74. *Cornbrash; Scarborough.* Leckenby Collection.

A. rustica [Walton MS.], J. Lycett, Supp. Mon. Moll. Gt. Ool. etc. (1863), pl. xl, f. 8, 8 *a*, p. 76. *Forest Marble; Farleigh.* Walton Collection.

A. subdentata, Roemer, W. Keeping, Foss. Upware and Brickhill (1883), pl. vi, f. 11, p. 122. *Lower Greensand; Upware.*

A. ungulata, J. Lycett, Supp. Mon. Moll. Gt. Ool. etc. (1863), pl. xxxv, f. 20, p. 72. *Cornbrash; Scarborough.* Leckenby Collection.

Astarte sp., W. Keeping, Foss. Upware and Brickhill (1883), pl. vi, f. 9 *a*, p. 122. *Lower Greensand; Upware.*

A. sp., W. Keeping, Foss. Upware and Brickhill (1883), pl. vi, f. 10 *a, b,* p. 122. *Lower Greensand; Upware.*

Aucella pallasi, Keyserling, A. Pavlow, Bull. Soc. Impér. Natural. Moscou. (1889), pl. iii, f. 4, 5. *Kimeridge Clay; Claxby.*

Avicula braamburiensis, Sowerby, J. Morris and J. Lycett, Mollusca Gt. Ool. part ii (1853), pl. xv, f. 7, p. 129. *Inferior Oolite (Grey Limestone); White Nab, near Scarborough.* Leckenby Collection.

A. cornueliana, d'Orbigny, W. Keeping, Foss. Upware and Brickhill (1883), pl. v, f. 2, p. 109. *Lower Greensand; Upware.*

A. cuneata, H. G. Seeley, A. M. N. H. ser. 3, XVII (1866), p. 179. *Red Chalk; Hunstanton.*

A. damnoniensis, v. Pterinæa damnoniensis.

A. danbyi, F. McCoy, A. M. N. H. ser. 2, VII (1851), p. 59, and Contrib. Brit. Pal. (1854), p. 194, and Brit. Palæoz. Foss. (1852), pl. i I, f. 11—15, p. 258. *Upper Ludlow; Benson Knot, Kendal.*

A. dubia, R. Etheridge, Appendix to W. H. Penning and A. J. Jukes-Browne, Geol. Cambridge (1881), pl. ii, f. 4, 4 *a*, p. 145. *Lower Chalk (Totternhoe Stone); Burwell.*

A. filata, R. Etheridge, Appendix to W. H. Penning and A. J. Jukes-Browne, Geol. Cambridge (1881), pl. ii, f. 3, 3 *a*, p. 144. *Lower Chalk (Totternhoe Stone); Burwell.*

A. pectinoides, v. Aviculopecten pectinoides.

A. subradiata, v. Pterinæa subradiata.

Aviculopecten cælatus (McCoy), F. McCoy, Brit. Palæoz. Foss. (1855), pl. iii E, f. 5, p. 483. R. Etheridge, jun., A. M. N. H. ser. 4, XVIII (1876), p. 98. *Carboniferous Limestone; Lowick.* Presented by the Rev. J. J. Jenkinson.

A. cancellatus (McCoy), F. McCoy, Brit. Palæoz. Foss. (1855), pl. iii E, f. 3, 3 *a*, p. 483. *Carboniferous Limestone; Lowick.* Presented by the Rev. J. J. Jenkinson.

Aviculopecten concavus (McCoy), F. McCoy, Brit. Palæoz. Foss. (1855), pl. iii E, f. 2, 2 *a*, p. 484. *Carboniferous Limestone; Lowick.* Presented by W. Hopkins, Esq.

A. docens, F. McCoy, Brit. Palæoz. Foss. (1855), pl. iii E, f. 6, 7, p. 485. *Carboniferous Limestone; f. 6 from Lowick, f. 7 from Derbyshire.* Presented by the Rev. J. J. Jenkinson.

A. pectinoides (Sowerby). *Avicula pectinoides*, J. de C. Sowerby, Trans. Geol. Soc. ser. 2, v (1840), pl. liv, f. 2. *Upper Devonian; Petherwin.*

A. planoradiatus, F. McCoy, A. M. N. H. ser. 2, vii (1851), p. 171, and Contrib. Brit. Pal. (1854), p. 204, and Brit. Palæoz. Foss. (1855), pl. iii E, f. 8, p. 489. *Carboniferous Limestone; Derbyshire.* Presented by W. Hopkins, Esq.

A. ruthveni, F. McCoy, A. M. N. H. ser. 2, vii (1851), p. 172, and Contrib. Brit. Pal. (1854), p. 204, and Brit. Palæoz. Foss. (1855), pl. iii E, f. 4, p. 489. *Carboniferous Limestone; Dent.* Presented by the Rev. J. J. Jenkinson.

A. segregatus (McCoy), F. McCoy, Brit. Palæoz. Foss. (1855), pl. iii E, f. 1, p. 489. *Carboniferous Limestone; Lowick.*

Axinæa proxima (Wood). *Pectunculus proximus*, S. V. Wood, Eoc. Bivalves i (1864), pl. xvii, f. 11, p. 99. *Bracklesham Beds; Huntingbridge.* Fisher Collection.

Cardiomorpha orbicularis, F. McCoy, A. M. N. H. ser. 2, xii (1853), p. 189, and Contrib. Brit. Pal. (1854), p. 257, and Brit. Palæoz. Foss. (1855), pl. iii i, f. 41, 41 *a*, p. 510. *Carboniferous Limestone; Derbyshire.* Presented by the Rev. J. J. Jenkinson.

Cardita rotundata?, Pictet and Roux, W. Keeping, Foss. Upware and Brickhill (1883), pl. vi, f. 7, p. 121. *Lower Greensand; Upware.*

Cardium buckmani, J. Lycett, Supp. Mon. Moll. Gt. Ool. etc. (1863), pl. xl, f. 18, 18 *a*. *Forest Marble; Laycock, Wilts.* Walton Collection.

C. cognatum, Phillips, J. Leckenby, Q. J. G. S. xv (1859), pl. iii, f. 8 *a, b*. *Kellaways Rock; Scarborough.* Leckenby Collection. Another specimen, J. Lycett, Supp. Mon. Moll. Gt. Ool. etc. (1863), pl. xxxvi, f. 3, p. 54. *Cornbrash; Scarborough.* Leckenby Collection.

Cardium crawfordi, J. Leckenby, Q. J. G. S. xv (1859), pl. iii, f. 9 *a, b,* p. 14. *Kellaways Rock; north side of the Castle Rock, Scarborough.* Leckenby Collection.

C. globosum, W. Bean, Mag. Nat. Hist. new series, III (1839), p. 60, woodcut 19. J. Lycett, Supp. Mon. Moll. Gt. Ool. etc. (1863), pl. xxxviii, f. 2, 2 *a, b,* p. 114. *Cornbrash; Scarborough.* Leckenby Collection.

C. pulsatum, G. F. Whidborne, Q. J. G. S. XXXIX (1883), pl. xviii, f. 9, 9 *a,* p. 525. *Inferior Oolite; Dundry.* Presented by the Rev. G. F. Whidborne, M.A.

C. purbeckense, P. de Loriol, Soc. Phys. et d'Hist. Nat. Genève, XVIII (1865), pl. iii, f. 15, 15 *a,* 16, 16 *a,* p. 45. *Purbeck Beds; Ridgway.* Fisher Collection.

C. subhillanum?, Leymerie, W. Keeping, Foss. Upware and Brickhill (1883), p. 119. *Lower Greensand; Upware.*

Cleidophorus? ovalis, F. McCoy, A. M. N. H. ser. 2, VII (1851), p. 55, and Contrib. Brit. Pal. (1854), p. 190, and Brit. Palæoz. Foss. (1852), pl. i K, f. 7, 8, p. 273. *Fig. 7 from Middle Bala; Dolydd Ceiriog Waterfall. Fig. 8 from the Denbighshire Grits; Plas Madoc, Llanwrst.*

C. planulatus (Conrad), F. McCoy, Brit. Palæoz. Foss. (1852), pl. i K, f. 9, p. 273. *Wenlock Beds; Golden Grove, near Llandeilo.*

Corbicella subæquilatera, J. Lycett, Supp. Mon. Moll. Gt. Ool. etc. (1863), pl. xxxv, f. 12, p. 69. *Cornbrash; Scarborough.* Leckenby Collection.

C. subangulata, J. Lycett, Supp. Mon. Moll. Gt. Ool. etc. (1863), pl. xl, f. 9, p. 70. *Forest Marble; Laycock.* Walton Collection.

Corbis gaultina (Pictet), A. J. Jukes-Browne, Q. J. G. S. XXXI (1875), pl. xv, f. 9, p. 300. *Cambridge Greensand; Cambridge.*

C. rotunda, [Walton MS.], J. Lycett, Supp. Mon. Moll. Gt. Ool. etc. (1863), pl. xl, f. 17, p. 60. *Great Oolite; Hampton Cliffs, near Bath.* Walton Collection.

Corbula hulliana, Morris, J. Lycett, Supp. Mon. Moll. Gt. Ool. etc. (1863), pl. xxxvii, f. 5, p. 64. *Forest Marble; Hinton.*

Ctenodonta elongata, H. Hicks, in Salter, Cat. (1873), p. 24. *Tremadoc Beds; Ramsey Island.* This specimen cannot be found.

C. hughesi, Salter, Cat. (1873), p. 82. *Lower Llandovery; Sefin Llettyrhyddod, near Llandovery.*

C. levata (Hall). *Nucula levata*, Hall, F. McCoy, Brit. Palæoz. Foss. (1852), pl. i κ, f. 4, 5, 5 *a*, p. 285. *Bala Beds; Dinas Bran, Llangollen.* The specimens cannot be identified with certainty.

C. menapiensis, Hicks. *C. rotunda*, H. Hicks, in Salter, Cat. (1873), p. 24. *Tremadoc Beds; Ramsey Island.* This specimen cannot be found.

C. poststriata (Emmons). *Nuculites poststriatus*, F. McCoy, Brit. Palæoz. Foss. (1852), pl. i κ, f. 6, p. 286. *Denbighshire Flags; Gwddelwern, Denbigh.*

C. rotunda, v. Ctenodonta menapiensis.

C. subæqualis, v. Arca subæqualis.

C. sp. 1, Salter, Cat. (1873), p. 152. *Wenlock Limestone; Dudley.* Fletcher Collection.

C. sp. 2, Salter, Cat. (1873), p. 152. *Wenlock Limestone; Dudley.* Fletcher Collection.

C. sp. 3, Salter, Cat. (1873), p. 152. *Wenlock Limestone; Dudley.* Fletcher Collection.

C. sp. 4, Salter, Cat. (1873), p. 152. *Wenlock Limestone; Dudley.* Fletcher Collection.

Cucullæa angusta, J. de C. Sowerby, Trans. Geol. Soc. ser. 2, v (1840), pl. liii, f. 25. *Upper Devonian; Marwood.*

C. cancellata, Phillips, J. Morris and J. Lycett, Mollusca Gt. Ool. part ii (1853), pl. xiv, f. 12, p. 132. *Inferior Oolite (Grey Limestone); Cloughton, near Scarborough.* Leckenby Collection.

C. clathrata, J. Leckenby, Q. J. G. S. xv (1859), pl. iii, f. 4, p. 13. *Kellaways Rock; Scarborough.* Leckenby Collection.

Another specimen, J. Lycett, Supp. Mon. Moll. Gt. Ool. etc. (1863), pl. xxxix, f. 4, 4 *a*, p. 44. *Cornbrash; Scarborough.* Leckenby Collection.

Cucullæa corallina, Damon, J. Lycett, Supp. Mon. Moll. Gt. Ool. etc. (1863), pl. xxxix, f. 3, p. 43. *Cornbrash; Scarborough.* Leckenby Collection.

C. donningtonensis, W. Keeping, Foss. Upware and Brickhill (1883), pl. viii, f. 9 *a, b,* p. 152. *Lower Greensand (derived); Upware.*

C. hardingi, J. de C. Sowerby, Trans. Geol. Soc. ser. 2, v (1840), pl. liii, f. 26, 27. *Upper Devonian; Marwood.*

C. minima, J. Leckenby, Q. J. G. S. xv (1859), pl. iii, f. 5, p. 13. *Kellaways Rock; the Castle Rock, Scarborough.* Leckenby Collection.

C. subnana (Pictet and Roux), W. Keeping, Foss. Upware and Brickhill (1883), pl. v, f. 10, p. 115. *Lower Greensand; Upware.*

C. trapezium, J. de C. Sowerby, Trans. Geol. Soc. ser. 2, v (1840), pl. liii, f. 24. *C. unilateralis*, J. de C. Sowerby, *ibid.* (1840), pl. liii, f. 23. *Upper Devonian; Marwood.*

C. unilateralis, v. Cucullæa trapezium.

C. vagans, W. Keeping, Foss. Upware and Brickhill (1883), pl. viii, f. 8 *a, b,* p. 151. *Lower Greensand (derived); Upware.*

C. sp.?, W. Keeping, Foss. Upware and Brickhill (1883), pl. v, f. 8, p. 115. *Lower Greensand; Upware.*

Cypricardia arcadiformis, W. Keeping, Foss. Upware and Brickhill (1883), pl. vi, f. 6, p. 120. *Lower Greensand; Upware.*

C. squamosa, W. Keeping, Foss. Upware and Brickhill (1883), pl. vi, f. 5 *a—c,* p. 120. *Lower Greensand; Upware.*

Cyprina bella, J. Lycett, Supp. Mon. Moll. Gt. Ool. etc. (1863), pl. xl, f. 15, 15 *a,* p. 71. *Forest Marble; Laycock, Wilts.* Walton Collection.

C. davidsoni, J. Lycett, Supp. Mon. Moll. Gt. Ool. etc. (1863), pl. xxxvi, f. 6, 6 *a,* p. 71. *Forest Marble; Farleigh.* Walton Collection.

C. obtusa, W. Keeping, Foss. Upware and Brickhill (1883), pl. vi, f. 13 *a—c,* p. 124. *Lower Greensand; Upware.*

C. sedgwicki (Walker), W. Keeping, Foss. Upware and Brickhill (1883), pl. vi, f. 12 *a—c,* p. 123. *Lower Greensand; Upware.*

Dolabra elliptica, F. McCoy, A. M. N. H. ser. 2, VII (1851), p. 52, and Contrib. Brit. Pal. (1854), p. 187, and Brit. Palæoz. Foss. (1852), pl. i L, f. 10, 10 *a*, p. 269. *Palæarca? elliptica*, Salter, Cat. (1873), p. 191. *Upper Ludlow Tilestones; Storm Hill, Llandeilo.*

D. obtusa, F. McCoy, A. M. N. H. ser. 2, VII (1851), p. 53, and Contrib. Brit. Pal. (1854), p. 188, and Brit. Palæoz. Foss. (1852), pl. i K, f. 30, p. 270. *Upper Ludlow Tilestones; Storm Hill, Llandeilo.*

Edmondia oblonga, v. Sanguinolites oblongus.

Edmondia rudis, F. McCoy, A. M. N. H. ser. 2, XII (1853), p. 190, and Contrib. Brit. Pal. (1854), p. 259, and Brit. Palæoz. Foss. (1855), pl. iii F, f. 9, p. 502. *Carboniferous Limestone; Lowick.*

E. scalaris (McCoy), F. McCoy, Brit. Palæoz. Foss. (1855), pl. iii H, f. 6, 6 *a*, p. 502. *Carboniferous Limestone; Lowick.*

Exogyra conica (Sowerby) **var.**, W. Keeping, Foss. Upware and Brickhill (1883), pl. iv, f. 3 *a*, *b*, p. 101. *Lower Greensand; Upware.*

E. davidsoni, G. F. Whidborne, Q. J. G. S. XXXIX (1883), pl. xv, f. 10, 10 *a*, p. 495. *Inferior Oolite; Frocester.* Presented by the Rev. G. F. Whidborne, M.A.

E. rauliniana, d'Orbigny, var. **arcula**, H. G. Seeley, A. M. N. H. ser. 3, XVII (1866), p. 177. *Red Chalk; Hunstanton.*

Gervillia acuta, Sowerby, J. Morris and J. Lycett, Mollusca Gt. Ool. part ii (1853), pl. xiv, f. 1, 1 *a*, p. 20. *Inferior Oolite (Grey Limestone); Scarborough.* Leckenby Collection.

G. intermedia, G. F. Whidborne, Q. J. G. S. XXXIX (1883), pl. xvi, f. 8, 9, p. 516. *Inferior Oolite; f. 8 from Frocester, f. 9 from Bradford Abbas.* Presented by the Rev. G. F. Whidborne.

G. tortuosa (Sowerby) **var.**, J. Lycett, Supp. Mon. Moll. Gt. Ool. etc. (1863), pl. xl, f. 25, p. 37. *Cornbrash; Scarborough.* Leckenby Collection. This specimen is missing.

G. waltoni, J. Lycett, Supp. Mon. Moll. Gt. Ool. etc. (1863), pl. xxxii, f. 4, 4 *a*, *b*, p. 110. *Forest Marble; Farleigh.* Walton Collection.

Glyptarca primæva, H. Hicks, Q. J. G. S. XXIX (1873), pl. v, f. 3, 3 *a*, p. 58. *Tremadoc Beds; Ramsey Island.* Presented by Dr H. Hicks.

Goniophora grandis, Salter, Cat. (1873), p. 151. *Wenlock Limestone; Dudley.* Fletcher Collection.

Gouldia ovalis (Quenstedt), G. F. Whidborne, Q. J. G. S. XXXIX (1883), pl. xviii, f. 17, 17 *a*, p. 528. *Inferior Oolite; Bradford Abbas.* Presented by the Rev. G. F. Whidborne.

Grammysia cingulata (Hisinger), **var. a**, F. McCoy, Brit. Palæoz. Foss. (1852), p. 280. *Wenlock Shale; Dudley: and Upper Ludlow; Benson Knot.*

    G. cingulata var. *obliqua*, F. McCoy, *ibid.* (1852), p. 280. *Upper Ludlow; High Thorns, Underbarrow.*

G. cingulata var. *triangulata*, v. Grammysia triangulata.

G. extrasulcata (Salter), F. McCoy, Brit. Palæoz. Foss. (1852), pl. i K, f. 29, p. 281. *Upper Ludlow; Benson Knot.*

G. rotundata (Sowerby), F. McCoy, Brit. Palæoz. Foss. (1852), pl. i K, f. 26, 27, p. 281. *Upper Ludlow; Benson Knot, Kendal.*

G. triangulata (Salter). *G. cingulata* var. *triangulata*, F. McCoy, Brit. Palæoz. Foss. (1852), pl. i K, f. 28, p. 280. *UpperLudlow; Benson Knot.*

Harpax parkinsoni (Quenstedt?), G. F. Whidborne, Q. J. G. S. XXXIX (1883), pl. xv, f. 20, p. 513. *Inferior Oolite; Yeovil Junction.* Presented by the Rev. G. F. Whidborne.

H. waltoni, J. Lycett, Supp. Mon. Moll. Gt. Ool. etc. (1863), pl. xxxii, f. 1, 1 *a, b*, p. 110. *Forest Marble; Farleigh.* Walton Collection.

Hinnites gradus (Bean), J. Lycett, Supp. Mon. Moll. Gt. Ool. (1863), pl. xxxiii, f. 10, 10 *a*, p. 35. *Cornbrash; Scarborough.* Leckenby Collection.

H. salteri, H. G. Seeley, A. M. N. H. ser. 3, XVII (1866), p. 178. *Red Chalk; Hunstanton.*

H. trilinearis, H. G. Seeley, A. M. N. H. ser. 3, VII (1861), pl. vi, f. 2, p. 119. *Cambridge Greensand; Cambridge.*

Hinnites trilinearis, Seeley, **var.**, H. G. Seeley, A. M. N. H. ser. 3, XVII (1866), p. 178. *Red Chalk ; Hunstanton.*

Inoceramus convexus, R. Etheridge, Appendix to W. H. Penning and A. J. Jukes-Browne, Geol. Cambridge (1881), pl. ii, f. 6, 6 *a*, p. 143. *Lower Chalk (Totternhoe Stone); Burwell.*

I. convexus, Etheridge, var. **quadratus**, R. Etheridge, Appendix to W. H. Penning and A. J. Jukes-Browne, Geol. Cambridge (1881), pl. ii, f. 7, p. 143. *Lower Chalk (Totternhoe Stone); Burwell.*

I. latus, Mantell, var. **reachensis**, R. Etheridge, Appendix to W. H. Penning and A. J. Jukes-Browne, Geol. Cambridge (1881), pl. i, f. 3, 3 *a*, p. 142. *Lower Chalk (Totternhoe Stone); Burwell.*

I. sulcatus, Parkinson. *Arca ? sulcata*, H. G. Seeley, A. M. N. H. ser. 3, VII (1861), pl. vi, f. 3, p. 119, and *ibid.* XVII (1866), p. 178. A. J. Jukes-Browne, Q. J. G. S. XXXI (1875), p. 298. *Cambridge Greensand ; Cambridge.*

Isoarca agassizi, Pictet and Roux, A. J. Jukes-Browne, Q. J. G. S. XXXI (1875), pl. xv, f. 1—3, p. 300. *Cambridge Greensand ; Cambridge.*

I. scarburgensis, J. Lycett, Supp. Mon. Moll. Gt. Ool. etc. (1863), pl. xxxix, f. 5, 5 *a*, p. 45. *Cornbrash ; Scarborough.* Leckenby Collection.

Isocardia ? clarissima [Bean MS.], J. Leckenby, Q. J. G. S. xv (1859), pl. iii, f. 10 *a*, *b*, p. 14. *Kellaways Rock ; north side of the Castle Rock, Scarborough.* Leckenby Collection.

I. cordata, Buckman, J. Morris and J. Lycett, Mollusca Gt. Ool. (1853), pl. xv, f. 5, p. 135. *Millepore Bed ; Stainton Dale Cliff, near Scarborough.* Leckenby Collection.

I. nitida, Phillips, J. Lycett, Supp. Mon. Moll. Gt. Ool. etc. (1863), pl. xxxviii, f. 6, 6 *a*, *b*, p. 57. *Cornbrash ; Scarborough.* Leckenby Collection.

I. tenera, Sowerby, J. Lycett, Supp. Mon. Moll. Gt. Ool. etc. (1863), pl. xxxviii, f. 5, 5 *a*, *b*, p. 57. *Cornbrash ; Scarborough.* Leckenby Collection.

Leda anglica, d'Orbigny, J. Lycett, Supp. Mon. Moll. Gt. Ool. etc. (1863), pl. xxxix, f. 7, p. 45. *Cornbrash; Scarborough.* Leckenby Collection.

L. caudata (Donovan), S. V. Wood, Crag Mollusca, part ii (1850), pl. x, f. 12 *a, b,* p. 92. *Bridlington Crag; Bridlington.* Leckenby Collection.

L. scapha (d'Orbigny), J. S. Gardner, Q. J. G. S. xl (1884), pl. v, f. 23, p. 138. *Speeton Clay; Speeton.* This specimen is missing.

L. seeleyi, J. S. Gardner, Q. J. G. S. xl (1884), pl. v, f. 17, 18, p. 137. *Speeton Clay; Speeton.* Leckenby Collection.

L. spathulata, Forbes, J. S. Gardner, Q. J. G. S. xl (1884), pl. v, f. 31, 32, p. 139 (f. 31 is missing). *Speeton Clay; Speeton.*

L. subrecurva (Phillips), J. S. Gardner, Q. J. G. S. xl (1884), pl. v, f. 24, 25, p. 135. *Speeton Clay; Speeton.* Leckenby Collection.

Leptodomus constrictus, F. McCoy, A. M. N. H. ser. 2, viii (1851), p. 486, and Contrib. Brit. Pal. (1854), p. 237, and Brit. Palæoz. Foss. (1855), pl. ii A, f. 10*, p. 396. *Upper Devonian; Marwood.*

L. costellatus, F. McCoy, A. M. N. H. ser. 2, vii (1851), p. 175, and Contrib. Brit. Pal. (1854), p. 207, and Brit. Palæoz. Foss. (1855), pl. iii F, f. 5, 5 *a,* p. 508. *Carboniferous Limestone; Craige, near Kilmarnock.*

L. globulosus, v. Orthonota globulosa.

L. truncatus, v. Orthonata truncata.

Lima amnifera, G. F. Whidborne, Rep. Brit. Assoc. (1882), p. 534, and Q. J. G. S. xxxix (1883), pl. xvii, f. 2, p. 504. *Inferior Oolite; Yeovil Junction.* Presented by the Rev. G. F. Whidborne, M.A.

L. echinata, R. Etheridge, Appendix to W. H. Penning and A. J. Jukes-Browne, Geol. Cambridge (1881), pl. ii, f. 2, 2 *a, b,* p. 144. *Lower Chalk (Totternhoe Stone); Burwell.*

L. farringdonensis, Sharpe, W. Keeping, Foss. Upware and Brickhill (1883), pl. v, f. 12 *a, b,* p. 112. *Lower Greensand; Upware.*

Lima helvetica, Oppel, J. Lycett, Supp. Mon. Moll. Gt. Ool. etc. (1863), pl. xxxiii, f. 8, 8 *a*, p. 41. *Cornbrash; Scarborough.* Leckenby Collection.

L. interlineata, A. J. Jukes-Browne, Q. J. G. S. XXXIII (1877), pl. xxi, f. 10, 10 *a*, p. 502. *Cambridge Greensand; Cambridge.*

L. longa, Roemer, W. Keeping, Foss. Upware and Brickhill (1883), pl. v, f. 6 *a, b*, p. 112. *Lower Greensand; Upware.*

L. ornata, R. Etheridge, Appendix to W. H. Penning and A. J. Jukes-Browne, Geol. Cambridge (1881), pl. iii, f. 2, 2 *a*, p. 144. *Cambridge Greensand; Cambridge.*

L. rauliniana, d'Orbigny, A. J. Jukes-Browne, Q. J. G. S. XXXIII (1877), pl. xxi, f. 9, p. 502. *Cambridge Greensand; Cambridge.*

Lithodomus jenkinsoni, F. McCoy, A. M. N. H. ser. 2, XII (1853), p. 189, and Contrib. Brit. Pal. (1854), p. 258, and Brit. Palæoz. Foss. (1855), pl. iii F, f. 2, 2 *a*, p. 493. *Carboniferous Limestone; Lowick.*

L. phosphaticus, W. Keeping, Foss. Upware and Brickhill (1883), pl. vii, f. 1 *a, b*, p. 127. *Lower Greensand; Potton.*

Lucina beani, J. Lycett, Supp. Mon. Moll. Gt. Ool. etc. (1863), pl. xxxviii, f. 3, p. 59. *Cornbrash; Scarborough.* Leckenby Collection. There is some doubt about this being the specimen figured.

L.? burtonensis, J. Lycett, Supp. Mon. Moll. Gt. Ool. etc. (1863), pl. xl, f. 20, 20 *a, b*, p. 59. *Forest Marble; Burton Bradstock.* Walton Collection.

L. columbella, Lamarck, S. V. Wood, Crag Mollusca, part ii (1850), p. 143. *Red Crag? locality?* These specimens cannot be found.

L. tenera (Sowerby), A. J. Jukes-Browne, Q. J. G. S. XXXI (1875), pl. xv, f. 10—12, p. 300. *Figs.* 10, 12, *Cambridge Greensand; Cambridge. Fig.* 11, *Gault; Folkestone.*

Lyrodesma plana?, Conrad, F. McCoy, Brit. Palæoz. Foss. (1855), pl. i K, f. 17, 17 *a*, p. 272. *Bala Beds; Yspatty Evan.* The specimen does not agree well with the figure.

Macrodon hirsonensis (d'Archiac) var. **rugosa**, J. Lycett, Supp. Mon. Moll. Gt. Ool. etc. (1863), pl. xxxvi, f. 9, p. 113. *Forest Marble; Farleigh.* Walton Collection.

Modiola arcadiformis, W. Keeping, Foss. Upware and Brickhill (1883), pl. vii, f. 2 *a, b,* p. 128. *Lower Greensand; Potton.*

M.? **consobrina**, S. V. Wood, Eocene Bivalves I (1864), pl. xix f. 17, p. 76. *Bracklesham Beds; Alum Bay, Isle of Wight.* Fisher Collection.

M. gibbosa, W. H. Hudleston, Proc. Geol. Assoc. v (1878), p. 489. *Coral Rag; Malton.*

M. obesa, W. Keeping, Foss. Upware and Brickhill (1883), pl. vi, f. 3 *a, b,* p. 117. *Lower Greensand; Upware.*

M. pedernalis?, Roemer, W. Keeping, Foss. Upware and Brickhill (1883), pl. vi, f. 2 *a, b,* p. 117. *Lower Greensand; Upware.*

M. sp., W. Keeping, Foss. Upware and Brickhill (1883), p. 118. *Lower Greensand; Upware.*

Modiolopsis gradata (Salter). *M. nilssoni* (Hisinger), F. McCoy, Brit. Palæoz. Foss. (1852), pl. i I, f. 21, p. 267. *Lower Ludlow; Park Lane, Llandeilo.*

M. homfrayi, H. Hicks, Q. J. G. S. xxix (1873), pl. v, f. 16, 17, p. 49. *Tremadoc Beds; Ramsey Island.* Presented by Dr H. Hicks.

M. inflata, F. McCoy, A. M. N. H. ser. 2, VII (1851), p. 58, and Contrib. Brit. Pal. (1854), p. 193, and Brit. Palæoz. Foss. (1852), pl. i I, f. 16, p. 266. *Middle Bala; Pen Cerrig, Serth.*

M. modiolaris (Conrad), F. McCoy, Brit. Palæoz. Foss. (1852), pl. i I, f. 17, 18, p. 267. *M. m'coyi*, J. W. Salter, Appendix to A. C. Ramsay, Geol. North Wales (1866), p. 346, and *ibid.* second edition (1881), p. 552. *Middle Bala; fig.* 17 *Cader Dinmael, fig.* 18 *Horderley.*

M. nilssoni, v. Modiolopsis gradata.

M. planata, Salter, Cat. (1873), p. 182. *Upper Ludlow; Kirkby Moor, Kendal.*

M. planata, var., Salter, Cat. (1873), p. 182. *Upper Ludlow; Benson Knot.*

Modiolopsis postlineata, F. McCoy, A. M. N. H. ser. 2, VII (1851), p. 58, and Contrib. Brit. Pal. (1854), p. 193, and Brit. Palæoz. Foss. (1852), pl. i I, f. 22, p. 268. *Middle Bala; Allt-yr-Anker, Meifod.*

M. solvensis, H. Hicks, Q. J. G. S. XXIX (1873), pl. v, f. 18, p. 50. *Tremadoc Beds ; Ramsey Island.* Presented by Dr H. Hicks.

Myacites modica (Bean), J. Lycett, Supp. Mon. Moll. Gt. Ool. etc. (1863), pl. xliii, f. 1, 1 a, p. 83. *Cornbrash ; Scarborough.* Leckenby Collection.

M. recurvum (Phillips), J. Lycett, Supp. Mon. Moll. Gt. Ool. etc. (1863), pl. xxxvi, f. 4, 4 a, p. 81. *Cornbrash; Scarborough.* Leckenby Collection.

M. sinistra (Agassiz), J. Lycett, Supp. Mon. Moll. Gt. Ool. etc. (1863), pl. xxxv, f. 17 a, b, p. 82. *Cornbrash; Scarborough.* Leckenby Collection.

M. subsidens, G. F. Whidborne, Q. J. G. S. XXXIX (1883), pl. xviii, f. 24, 24 a, p. 535. *Inferior Oolite ; Dundry.* Presented by the Rev. G. F. Whidborne, M.A.

Myoconcha implana, G. F. Whidborne, Rep. Brit. Assoc. (1882), p. 534, and Q. J. G. S. XXXIX (1883), pl. xviii, f. 22, pl. xix, f. 5, p. 530. *Inferior Oolite ; Bradford Abbas.* Presented by the Rev. G. F. Whidborne.

Mytilus (Modiola) cuneatus, Sowerby, J. Morris and J. Lycett, Mollusca Gt. Ool. part ii (1853), pl. xiv, f. 8, p. 131. *Millepore Bed; Stainton Dale Cliffs, near Scarborough.* Leckenby Collection.

M. (Modiola) leckenbyi, J. Morris and J. Lycett, Mollusca Gt. Ool. part ii (1853), pl. xiv, f. 9, p. 131. *Millepore Bed ; Stainton Dale Cliffs, near Scarborough.* Leckenby Collection.

Mytilus minimus, Salter, Cat. (1873), p. 182. *Upper Ludlow ; Lesmahagow.*

Nucula bivirgata, Sowerby, A. J. Jukes-Browne, Q. J. G. S. XXXI (1875), pl. xv, f. 6, p. 299. *Cambridge Greensand; Cambridge.*

N. cornueliana, d'Orbigny, J. S. Gardner, Q. J. G. S. XL (1884), pl. v, f. 5, 6, p. 129. *Lower Greensand ; Benniworth.*

N. levata, v. Ctenodonta levata.

Nucula planata, Deshayes, J. S. Gardner, Q. J. G. S. XL (1884), pl. v, f. 1, 2, 3, p. 126. *Fig.* 1, *Lower Greensand; Atherfield. Figs.* 2 *and* 3, *Speeton Clay; Speeton.* Leckenby Collection.

N. pullastriformis, F. McCoy, Brit. Palæoz. Foss. (1852), p. 397. *Pullastra antiqua*, J. de C. Sowerby, Trans. Geol. Soc. ser. 2, v (1840), pl. liii, f. 28. *Upper Devonian; Marwood.*

N. rhomboidea, H. G. Seeley, A. M. N. H. ser. 3, VII (1861), pl. vi, f. 5, p. 120. *Cambridge Greensand; Cambridge.*

N. subelliptica, H. G. Seeley, A. M. N. H. ser. 3, VII (1861), pl. vi, f. 4, p. 120. *Cambridge Greensand; Cambridge.* Presented by J. Carter, Esq., F.G.S.

Nuculites poststriatus, v. Ctenodonta poststriata.

Opis leckenbyi, Wright, J. Lycett, Supp. Mon. Moll. Gt. Ool. etc. (1863), pl. xxxvii, f. 9, 9 *a*, p. 61. *Cornbrash; Scarborough.* Leckenby Collection.

O. luciensis, d'Orbigny, J. Lycett, Supp. Mon. Moll. Gt. Ool. etc. (1863), pl. xl, f. 19, 19 *a*, p. 62. *Great Oolite; Box Tunnel, near Bath.* Walton Collection.

O. neocomiensis, d'Orbigny, W. Keeping, Foss. Upware and Brickhill (1883), pl. vi, f. 8 *a—c*, p. 121. *Lower Greensand; Upware.*

Orthonota affinis (McCoy). *Tellinites affinis*, F. McCoy, A. M. N. H. ser. 2, VII (1851), p. 51, and Contrib. Brit. Pal. (1854), p. 186, and Brit. Palæoz. Foss. (1852), pl. i K, f. 31, p. 286. *Upper Ludlow; Benson Knot, Kendal.*

O. angulifera (McCoy). *Sanguinolites anguliferus*, F. McCoy, A. M. N. H. ser. 2, VII (1851), p. 56, and Contrib. Brit. Pal. (1854), p. 191, and Brit. Palæoz. Foss. (1852), pl. i K, f. 19, 20 p. 276. *Upper Ludlow; Benson Knot, Kendal.*

O. decipiens (McCoy). *Sanguinolites decipiens*, F. McCoy, Brit. Palæoz. Foss. (1852), pl. i I, f. 24, p. 277. *Upper Ludlow; Benson Knot.*

O. globulosa (McCoy). *Leptodomus globulosus*, F. McCoy, A. M. N. H. ser. 2, VII (1851), p. 57, and Contrib. Brit. Pal. (1854), p. 192, and Brit. Palæoz. Foss. (1852), pl. i K, f. 11, 11 *a*, p. 278. *Upper Ludlow; Tenterfell, Kendal.*

Orthonota nasuta (Conrad), F. McCoy, Brit. Palæoz. Foss. (1852), pl. i i, f. 23, p. 275. *Middle Bala; Horderley West.*

O. semisulcata (Sowerby), F. McCoy, Brit. Palæoz. Foss. (1852), pl. i к, f. 25, p. 275. *Upper Ludlow; Kirkby Moor, Kendal.*

O. truncata (McCoy). *Leptodomus truncatus*, F. McCoy, A. M. N. H. ser. 2, vii (1851), p. 57, and Contrib. Brit. Pal. (1854), p. 192, and Brit. Palæoz. Foss. (1852), pl. i к, f. 21—24, p. 279. *Upper Ludlow; Benson Knot, Kendal.*

Ostrea acutirostris, Nilsson, R. Etheridge, Appendix to W. H. Penning and A. J. Jukes-Browne, Geol. Cambridge (1881), pl. iii, f. 5, 6, p. 146. *Lower Chalk (Totternhoe Stone); Burwell.*

O. (Exogyra) couloni (Defrance), W. Keeping, Foss. Upware and Brickhill (1883), p. 100. *Lower Greensand; Upware.*

O. curvirostris, Nilsson, var. **inflexa**, R. Etheridge, Appendix to W. H. Penning and A. J. Jukes-Browne, Geol. Cambridge, (1881), pl. iii, f. 7, 8, p. 146. *Lower Chalk (Totternhoe Stone); Burwell.*

O. (Gryphea) dilatata?, Sowerby, W. Keeping, Foss. Upware and Brickhill (1883), p. 102. *Lower Greensand; Upware.*

O. frons, Parkinson, var. **macroptera**, Sowerby, W. Keeping, Foss. Upware and Brickhill (1883), p. 102. *Lower Greensand; Upware.*

O. knorri, Voltz, G. F. Whidborne, Q. J. G. S. xxxix (1883), pl. xv, f. 2, 3, 3 *a*, p. 490. *Fig. 2, Inferior Oolite; Bradford Abbas. Fig. 3, Fuller's Earth; Frome.* Presented by the Rev. G. F. Whidborne, M.A.

O. lagena, H. G. Seeley, A. M. N. H. ser. 3, vii (1861), pl. v, f. 2, p. 117. *Cambridge Greensand; Cambridge.*

O. (Exogyra) lingulata [Walton MS.], J. Lycett, Supp. Mon. Moll. Gt. Ool. etc. (1863), pl. xxxii, f. 2, 2 *a, b*, p. 108. *Forest Marble; Farleigh and Pound Hill.* Walton Collection.

O. marshi, Sowerby, J. Morris and J. Lycett, Mollusca Gt. Ool. part ii (1853), pl. xiv, f. 2, p. 126. *Inferior Oolite (Grey Limestone); Gristhorpe Bay, near Scarborough.*

Another specimen, young individual, Morris and Lycett,

ibid. pl. xiv, f. 2 *a*, p. 127. *Inferior Oolite (Whitwell Lime-stone); near Whitwell, Yorkshire.* Leckenby Collection.

Ostrea sphæroidalis, G. F. Whidborne, Rep. Brit. Assoc. (1882), p. 534, and Q. J. G. S. xxxix (1883), pl. xv, f. 5, 6, p. 493. *Inferior Oolite; Yeovil Junction.* Presented by the Rev. G. F. Whidborne, M.A.

O. walkeri, W. Keeping, Foss. Upware and Brickhill (1883), pl. iv, f. 4 *a, b,* p. 103. *Lower Greensand; Upware.*

O. wiltonensis, J. Lycett, Supp. Mon. Moll. Gt. Ool. etc. (1863), pl. xxxiv, f. 1, 1 *a*, p. 108. *Forest Marble; Pound Hill.* Walton Collection.

O. zonulata, S. V. Wood, Eocene Bivalves i (1861), pl. x, f. 4 *a, b, c,* p. 34. *Bracklesham Beds; Hill Head, near Stubbington.* Fisher Collection.

Palæarca diagona, v. Anodontopsis quadrata.

P.? elliptica, v. Dolabra elliptica.

Palæarca oboloidea, H. Hicks, Q. J. G. S. xxix (1873), pl. v, f. 10, p. 48. *Tremadoc Beds; Ramsey Island.* Presented by Dr H. Hicks, F.R.S.

Pecten anisopleurus, Buvignier, J. Lycett, Supp. Mon. Moll. Gt. Ool. etc. (1863), pl. xxxiii, f. 5, 5 *a*, p. 34. *Cornbrash; Scarborough.* Leckenby Collection.

P. articulatus, Schlotheim, J. Lycett, Supp. Mon. Moll. Gt. Ool. etc. (1863), pl. xxxiii, f. 12, p. 32. *Cornbrash; Scarborough.* Leckenby Collection.

P. barretti, H. G. Seeley, A. M. N. H. ser. 3, vii (1861), pl. vi, f. 1, p. 118. A. J. Jukes-Browne, Q. J. G. S. xxxiii (1877), p. 500. *Cambridge Greensand; Cambridge.*

P. demissus, Phillips, J. Morris and J. Lycett, Mollusca Gt. Ool. (1853), pl. xiv, f. 7, p. 127. *Millepore Bed; Stainton Dale Cliffs, near Scarborough.* Leckenby Collection.

P. demissus, Phillips, var. **inutilis**, G. F. Whidborne, Q. J. G. S. xxxix (1883), pl. xv, f. 15, p. 498. *Inferior Oolite; Yeovil Junction.* Presented by the Rev. G. F. Whidborne.

Pecten dutemplei, d'Orbigny, W. Keeping, Foss. Upware and Brickhill (1883), p. 105. *Lower Greensand ; Upware.*

P. fenestralis, G. F. Whidborne, Rep. Brit. Assoc. (1882), p. 534, and Q. J. G. S. xxxix (1883), pl. xv, f. 12, 12 *a*, p. 500. *Inferior Oolite; Bradford Abbas.* Presented by the Rev. G. F. Whidborne.

P. fissicosta, R. Etheridge, Appendix to W. H. Penning and A. J. Jukes-Browne, Geol. Cambridge (1881), pl. ii, f. 1, 1 *a*, pl. iii, f. 1, 1 *a*, p. 141. *Lower Chalk (Totternhoe Stone); Burwell.*

P. inæquicostatus, Phillips, J. Lycett, Supp. Mon. Moll. Gt. Ool. etc. (1863), pl. xxxiii, f. 1 *a*, p. 32. *Coralline Oolite; Malton.* Leckenby Collection.

P. intermittens, G. F. Whidborne, Q. J. G. S. xxxix (1883), pl. xv, f. 13, 13 *a*, p. 500. *Inferior Oolite; Birdlip.* Presented by the Rev. G. F. Whidborne.

P. marullensis, Leymerie, W. Keeping, Foss. Upware and Brickhill (1883), pl. v, f. 11, p. 116. *Lower Greensand; Upware.*

P. michelensis, Buvignier, J. Lycett, Supp. Mon. Moll. Gt. Ool. etc. (1863), pl. xxxiii, f. 3, p. 34. *Cornbrash ; Scarborough.* Leckenby Collection.

P. orbicularis, Sowerby, var. **magnus**, W. Keeping, Foss. Upware and Brickhill (1883), pl. v, f. 1, p. 106. *Lower Greensand; Upware.*

P. (Neithea) ornithopus, W. Keeping, Foss. Upware and Brickhill (1883), pl. iv, f. 5 *a, b*, p. 107. *Lower Greensand ; Upware.*

P. raulinianus, d'Orbigny, W. Keeping, Foss. Upware and Brickhill (1883), p. 104. *Lower Greensand ; Upware.*

P. raulinianus ?, d'Orbigny, A. J. Jukes-Browne, Q. J. G. S. xxxiii (1877), p. 501. *Cambridge Greensand ; Cambridge.*

P. subspinosus, Schlotheim, J. Lycett, Supp. Mon. Moll. Gt. Ool. etc. (1863), pl. xl, f. 14, p. 113. *Forest Marble; Farleigh.* Walton Collection.

Pectunculus obliquus, W. Keeping, Foss. Upware and Brickhill (1883), pl. vi, f. 1 *a—d*, p. 116. *Lower Greensand ; Upware.*

P. *proximus*, v. Axinæa proxima.

Pectunculus sublævis, Sowerby, W. Keeping, Foss. Upware and Brickhill (1883), pl. v, f. 9 *a, b, c,* p. 115. *Lower Greensand; Upware.*

Perna lanceolata, Geinitz, H. G. Seeley, A. M. N. H. ser. 3, VII (1861), p. 122. *Cambridge Greensand; Cambridge.* Presented by J. Carter, Esq., F.G.S.

P. lissa, H. G. Seeley, A. M. N. H. ser. 3, XVII (1866), p. 178. *Red Chalk; Hunstanton.*

P. mytiloides, Lamarck, J. Lycett, Supp. Mon. Moll. Gt. Ool. etc. (1863), pl. xxxii, f. 3, p. 112. *Forest Marble; Farleigh.* Walton Collection.

P. obliqua [Walton MS.], J. Lycett, Supp. Mon. Moll. Gt. Ool. etc. (1863), pl. xxxiv, f. 2, 2 *a,* p. 112. *Forest Marble; Gustard, Wiltshire.* Walton Collection.

P. oblonga, H. G. Seeley, A. M. N. H. ser. 3, VII (1861), pl. vi, f. 6, p. 121. *Cambridge Greensand; Cambridge.*

P. rugosa, Goldfuss, **var.**, J. Morris and J. Lycett, Mollusca Gt. Ool. part ii (1853), pl. xiv, f. 16, p. 128. *Grey Limestone; Cloughton Wyke, near Scarborough.* Leckenby Collection.

P. semielliptica, H. G. Seeley, A. M. N. H. ser. 3, VII (1861), pl. vi, f. 7, p. 121. *Cambridge Greensand; Cambridge.* Presented by J. Carter, Esq., F.G.S.

P. transversa, H. G. Seeley, A. M. N. H. ser. 3, XVII (1866), p. 179. *Red Chalk; Hunstanton.*

Pholadomya bellula, G. F. Whidborne, Q. J. G. S. XXXIX (1883), pl. xix, f. 10, 10 *a,* p. 534. *Inferior Oolite; Bradford Abbas.* Presented by the Rev. G. F. Whidborne.

P. callæa, G. F. Whidborne, Q. J. G. S. XXXIX (1883), pl. xix, f. 7, 7 *a,* p. 532. *Inferior Oolite; Dundry.* Presented by the Rev. G. F. Whidborne, M.A.

P. decussata (Mantell) var. **triangulata**, H. G. Seeley, A. M. N. H. ser. 3, VII (1861), p. 122. *Cambridge Greensand; Cambridge.*

P. fortis, G. F. Whidborne, Q. J. G. S. XXXIX (1883), pl. xix, f. 8, p. 533. *Inferior Oolite; Dundry.* Presented by the Rev. G. F. Whidborne.

Pholadomya heraulti, Agassiz, J. Morris and J. Lycett, Mollusca Gt. Ool. part ii (1853), pl. xv, f. 4, p. 124. *Millepore Bed; Stainton Dale Cliffs.* Leckenby Collection.

P. newtoni, G. F. Whidborne, Q. J. G. S. xxxix (1883), pl. xix, f. 9, 9 *a*, p. 533. *Inferior Oolite; Bradford Abbas.* Presented by the Rev. G. F. Whidborne.

P. ovulum, Agassiz, J. Lycett, Supp. Mon. Moll. Gt. Ool. etc. (1863), pl. xxxv, f. 18, 18 *a*, p. 84. *Cornbrash; Yorkshire coast.* Leckenby Collection.

P. sæmanni, J. Morris and J. Lycett, Mollusca Gt. Ool. part ii (1853), pl. xv, f. 3, p. 123. *Millepore Bed; Gristhorpe.* Leckenby Collection.

Pholas costellata, J. Morris and J. Lycett, Mollusca Gt. Ool. part ii (1853), pl. xiii, f. 18, p. 142. *Millepore Bed; Stainton Dale Cliffs.* Leckenby Collection.

Pinna dundriensis, G. F. Whidborne, Q. J. G. S. xxxix (1883), pl. xvi, f. 10, p. 518. *Inferior Oolite; Dundry.* Presented by the Rev. G. F. Whidborne.

P. flexicostata, F. McCoy, Brit. Palæoz. Foss. (1855), pl. iii E, f. 11—13, p. 499. *Carboniferous Limestone; Lowick.*

P. spatula, F. McCoy, A. M. N. H. ser. 2, xii (1853), p. 188, and Contrib. Brit. Pal. (1854), p. 257, and Brit. Palæoz. Foss. (1855), pl. iii E, f. 9, 9 *a*, *b*, 10, p. 499. *Carboniferous Limestone; fig. 9 Derbyshire, fig. 10 Lowick.*

P. tegulata, R. Etheridge, Appendix to W. H. Penning and A. J. Jukes-Browne, Geol. Cambridge (1881), pl. i, f. 2, p. 142. *Lower Chalk (Totternhoe Stone); Burwell.*

Placuna sagittalis, G. F. Whidborne, Q. J. G. S. xxxix (1883), pl. xv, f. 17, p. 497. *Inferior Oolite; Yeovil Junction.* Presented by the Rev. G. F. Whidborne.

Plicatula carteroniana, d'Orbigny, W. Keeping, Foss. Upware and Brickhill (1883), pl. v, f. 4 *a*, *b*, p. 110. *Lower Greensand; Upware.*

P. equicostata, W. Keeping, Foss. Upware and Brickhill (1883), pl. v, f. 5 *a*, *b*, *c*, p. 111. *Lower Greensand; Upware.*

Plicatula minuta, H. G. Seeley, A. M. N. H. ser. 3, xvii (1866), p. 176. *Red Chalk; Hunstanton.*

P. sollasi, G. F. Whidborne, Q. J. G. S. xxxix (1883), pl. xv, f. 21, 21 *a*, 22, 22 *a*, *b*, p. 515. *Inferior Oolite; Dundry.* Presented by the Rev. G. F. Whidborne.

P. subserrata (Münster), G. F. Whidborne, Q. J. G. S. xxxix (1883), pl. xvi, f. 4, 5, p. 515. *Inferior Oolite; Bradford Abbas.* Presented by the Rev. G. F. Whidborne.

Pseudaxinus securiformis, v. Anodontopsis securiformis.

Pterinæa? asperula, F. McCoy, A. M. N. H. ser. 2, vii (1851), p. 60, and Contrib. Brit. Pal. (1854), p. 195, and Brit. Palæoz. Foss. (1852), pl. i i, f. 5, 5 *a*, p. 259. *Wenlock Shale; Builth Bridge.*

P. condor, Salter, Cat. (1873), p. 169. *Lower Ludlow; Dudley.* Fletcher Collection.

P. damnoniensis (Sowerby). *Avicula damnoniensis*, J. de C. Sowerby, Trans. Geol. Soc. ser. 2, v (1840), pl. liii, f. 22. *Upper Devonian; Marwood.*

P. demissa (Conrad), F. McCoy, Brit. Palæoz. Foss. (1852), pl. i i, f. 7, 8, p. 260. *Upper Ludlow; Benson Knot.*

P. exasperata, Salter, Cat. (1873), p. 150. *Wenlock Limestone; Dudley.* Fletcher Collection.

P. hians, F. McCoy, A. M. N. H. ser. 2, vii (1851), p. 60, and Contrib. Brit. Pal. (1854), p. 195, and Brit. Palæoz. Foss. (1852), pl. i i, f. 6, 6 *a*, p. 260. *P. retroflexa* var. *hians*, Salter, Cat. (1873), p. 169. *Aymestry Limestone; Aymestry.*

P. megaloba, F. McCoy, A. M. N. H. ser. 2, vii (1851), p. 61, and Contrib. Brit. Pal. (1854), p. 196, and Brit. Palæoz. Foss. (1852), pl. i i, f. 19, 20, p. 261. *Upper Ludlow Tilestones; Storm Hill, Llandeilo.*

P. pleuroptera (Conrad), F. McCoy, Brit. Palæoz. Foss. (1852), pl. i i, f. 1, 2, 2 *a*, *b*, p. 261. *Fig.* 1, *Lower Ludlow; Park Lane, near Llandeilo. Fig.* 2, *Bala Beds; Cyrn-y-Brain.*

P. retroflexa (Wahlenberg) var. **naviformis** (Conrad), F. McCoy, Brit. Palæoz. Foss. (1852), pl. i i, f. 9, 10, p. 262. *Upper Ludlow; Kirkby Moor, near Kendal.*

Pterinæa retroflexa (Wahlenberg) **var. a,** F. McCoy, Brit. Palæoz. Foss. (1852), p. 262. *Aymestry Limestone; Leintwardine.*

P. retroflexa (Wahlenberg) **var.,** (*cf.* Avicula erecta, Conrad), F. McCoy, Brit. Palæoz. Foss. (1852), p. 262. *Upper Ludlow; Laverock Lane, Kendal.*

P. retroflexa (Wahlenberg) **var.,** (with coarse imbrications), Salter, Cat. (1873), p. 150. *Wenlock Limestone; Dudley.* Fletcher Collection.

P. sowerbyi, F. McCoy, Brit. Palæoz. Foss. (1852), p. 263. *Aymestry Limestone; Leintwardine.*

P. subfalcata (Conrad), F. McCoy, Brit. Palæoz. Foss. (1852), pl. i I, f. 3, 3 *a,* p. 263. *Coniston Grit; Howgill Fell, near Sedbergh.*

P. subradiata (Sowerby). *Avicula subradiata,* J. de C. Sowerby. Trans. Geol. Soc. ser. 2, v (1840), pl. liv, f. 1. *Upper Devonian; South Petherwin.*

P. tenuistriata, F. McCoy, A. M. N. H. ser. 2, VII (1851), p. 62, and Contrib. Brit. Pal. (1854), p. 197, and Brit. Palæoz. Foss. (1852), pl. i I, f. 4, 4 *a,* p. 263. *Upper Ludlow; Benson Knot, Kendal.*

P. sp., (*cf.* lineatula), Salter, Cat. (1873), p. 150. *Wenlock Shale (or Lower Ludlow); Dudley.* Fletcher Collection.

Pteronites persulcatus, F. McCoy, A. M. N. H. ser. 2, VII (1851), p. 170, and Contrib. Brit. Pal. (1854), p. 202, and Brit. Palæoz. Foss. (1855), pl. iii F, f. 1, p. 480. *Carboniferous Limestone; Lowick.*

Pteroperna plana, J. Morris and J. Lycett, Mollusca Gt. Ool. part ii (1853), pl. xiv, f. 4, p. 128. *Inferior Oolite (Grey Limestone); Cloughton, near Scarborough.* Leckenby Collection.

Pullastra antiqua, v. Nucula pullastriformis.

Quenstedtia lævigata (Phillips), J. Morris and J. Lycett, Mollusca Gt. Ool. part ii (1853), pl. xiv, f. 13, p. 135. *Grey Limestone; Cloughton Wyke, near Scarborough.* Leckenby Collection.

Quenstedtia lævigata (Phillips) var. **elongata**, W. H. Hudleston, Proc. Geol. Assoc. v (1878), pl. iv, f. 11 *a*, *b*, p. 476. *Coralline Oolite; Malton.* Leckenby Collection.

Sanguinolites anguliferus, v. Orthonota angulifera.

Sanguinolites clava, F. McCoy, A. M. N. H. ser. 2, VII (1851), p. 172, and Contrib. Brit. Pal. (1854), p. 205, and Brit. Palæoz. Foss. (1855), pl. iii F, f. 12, 12 *a*, p. 504. *Carboniferous Limestone; Llangollen.*

S. decipiens, v. Orthonota decipiens.

S. iridinoides, F. McCoy, Brit. Palæoz. Foss. (1855), pl. iii F, f. 11, 11 *a*, p. 504. *Carboniferous Limestone; Lowick.*

S. oblongus (Portlock). *Edmondia oblonga*, F. McCoy, Brit. Palæoz. Foss. (1855), pl. iii F, f. 10, 10 *a*, p. 501. *Carboniferous Limestone; Lowick.*

S. subcarinatus, F. McCoy, A. M. N. H. ser. 2, VII (1851), p. 173, and Contrib. Brit. Pal. (1854), p. 205, and Brit. Palæoz. Foss. (1855), pl. iii F, f. 4, p. 506. *Carboniferous Limestone; Lowick.*

S. variabilis, F. McCoy, A. M. N. H. ser. 2, VII (1851), p. 174, and Contrib. Brit. Pal. (1854), p. 206, and Brit. Palæoz. Foss. (1855), pl. iii F, f. 6, 6 *a*, 7, 7 *a*, 8, p. 508. *Carboniferous Limestone; Lowick.*

Solemya primæva (Phillips). *Solenimya? primæva*, F. McCoy, Brit. Palæoz. Foss. (1855), pl. iii F, f. 3, 3 *a*, p. 519. *Carboniferous Limestone; Lowick.*

S. woodwardiana, J. Leckenby, Q. J. G. S. xv (1859), pl. iii, f. 7, p. 14. *Kellaways Rock; south side of the Castle Rock, Scarborough.* Leckenby Collection.

Solenimya? primæva, v. Solemya primæva.

Spondylus æquicostatus, R. Etheridge, Appendix to W. H. Penning and A. J. Jukes-Browne, Geol. Cambridge (1881), pl. ii, f. 5, p. 145. *Lower Chalk; Cherryhinton.*

S. gibbosus, d'Orbigny, H. G. Seeley, A. M. N. H. ser. 3, XVII (1866), p. 177. *Red Chalk; Hunstanton.*

Tancredia gibbosa, J. Lycett, Supp. Mon. Moll. Gt. Ool. etc. (1863), pl. xxxvi, f. 11, p. 68. *Forest Marble; Laycock.* Walton Collection.

Tellinites affinis, v. Orthonota affinis.

Tellinomya lingulæcomes, v. Lingulella davisi.

Thracia studeri, Agassiz, G. F. Whidborne, Q. J. G. S. xxxix (1883), pl. xix, f. 6, p. 531. *Inferior Oolite; Bradford Abbas.* Presented by the Rev. G. F. Whidborne, M.A.

Thracia? (**or Tellina**) **sp.**, W. Keeping, Foss. Upware and Brickhill (1883), pl. vi, f. 14, p. 126. *Lower Greensand; Upware.*

Trigonia arata, J. Lycett, Supp. Mon. Moll. Gt. Ool. etc. (1863), pl. xl, f. 2, p. 52. *Forest Marble; Farleigh.* Walton Collection.

T. bathonica, J. Lycett, Supp. Mon. Moll. Gt. Ool. etc. (1863), pl. xl, f. 3, p. 52, and Brit. Foss. Trig. (1872), pl. i, f. 3, p. 17. *Great Oolite; block below Hampton Cliffs.* Walton Collection.

T. clytia, d'Orbigny, J. Lycett, Supp. Mon. Moll. Gt. Ool. etc. (1863), pl. xl, f. 5, 5 *a*, p. 51. *Great Oolite; Box Tunnel, near Bath.* Walton Collection.

T. decorata, v. Trigonia signata.

T. hunstantonensis, v. Trigonia scapha.

T. keepingi, J. Lycett, Brit. Foss. Trig. (1877), pl. xxxv, f. 1, 2, p. 196. *Lower Greensand (Tealby Series); Claxby.*

T. leckenbyi, J. Lycett, Brit. Foss. Trig. (1874), pl. xvi, f. 1, p. 71. *Supra-Liassic Sands; Robin Hood's Bay, Yorkshire.* Leckenby Collection.

T. perlata, Agassiz, J. Lycett. Brit. Foss. Trig. (1872), pl. iii, f. 2, p. 22. *Coralline Oolite; Pickering.* Leckenby Collection.

T. rupellensis, d'Orbigny, J. Lycett, Brit. Foss. Trig. (1872), pl. viii, f. 4, p. 28. *T. clavellata*, var., J. Leckenby, Q. J. G. S. xv (1859), p. 8. *Kellaways Rock; Scarborough.* Leckenby Collection.

T. scapha, Agassiz, J. Lycett, Brit. Foss. Trig. (1877), pl. xxxviii, f. 6, p. 183. *T. hunstantonensis*, H. G. Seeley, A. M. N. H. ser. 3, xiv (1864), p. 276. *Lower Greensand; Norfolk.*

T. signata, Agassiz. *T. decorata*, Lycett, J. Morris and J. Lycett, Mollusca Gt. Ool. part ii (1853), pl. xv, f. 1, p. 133. *Inferior Oolite (Grey Limestone); Cloughton Wyke, near Scarborough.* Leckenby Collection.

Trigonia tealbyensis, J. Lycett, Brit. Foss. Trig. (1875), pl. xxviii, f. 7, p. 114. *Lower Greensand; Claxby.*

T. tripartita, Forbes, J. Lycett, Supp. Mon. Moll. Gt. Ool. etc. (1863), pl. xl, f. 4, p. 51. *Cornbrash; Chippenham.* Walton Collection.

T. woodwardi, J. Lycett, Brit. Foss. Trig. (1872), p. 40, woodcut. *Kimeridge Clay; Villersville, near Honfleur.* Leckenby Collection.

Unicardium depressum (Phillips), J. Morris and J. Lycett, Mollusca Gt. Ool. part ii (1853), pl. xiv, f. 10, p. 133. *Inferior Oolite (Grey Limestone); Cloughton Wyke, near Scarborough.* Leckenby Collection.

U. gibbosum, J. Morris and J. Lycett, Mollusca Gt. Ool. (1853), pl. xiv, f. 11, p. 132. *Millepore Bed; Stainton Dale Cliffs, near Scarborough.* Leckenby Collection.

U. sulcatum [Bean MS.], J. Leckenby, Q. J. G. S. xv (1859), pl. iii, f. 11, p. 14. *Kellaways Rock; south side of the Castle Rock, Scarborough.* Leckenby Collection.

GASTEROPODA.

Acmœa tenuistriata, v. Tectura tenuistriata.

Acroculia columbina, v. Capulus columbinus.

Acroculia euomphaloides (McCoy). *Capulus? euomphaloides,* F. McCoy, A. M. N. H. ser. 2, x (1852), p. 190, and Contrib. Brit. Pal. (1854), p. 242, and Brit. Palæoz. Foss. (1852), pl. i к, f. 39, p. 290. *Lower Ludlow; Green Quarry, Leintwardine.*

A. multiplicata, v. Capulus multiplicatus.

A. procœva, v. Capulus compressus.

Actæon retusus, Phillips, W. H. Hudleston, G. M. dec. 2, VIII (1881), pl. iv, f. 7, p. 127. *Coral Rag; Malton.* (Ayton, according to Mr Hudleston.) Leckenby Collection.

Actæonina cinerea, W. H. Hudleston, G. M. dec. 3, II (1885), pl. v, f. 8, 8 *a*, p. 206. *Inferior Oolite (Scarborough Limestone); Cloughton Wyke.* Leckenby Collection.

Actæonina fasciata, J. Lycett, Supp. Mon. Moll. Gt. Ool. etc. (1863), pl. xliv, f. 15, p. 107. *Forest Marble; Laycock.* Walton Collection.

A. humeralis (Phillips), W. H. Hudleston, G. M. dec. 3, II (1885), p. 202. *Dogger Series; Blue Wyke.* Leckenby Collection.

A. luidi (Lhwyd), J. Lycett, Supp. Mon. Moll. Gt. Ool. etc. (1863), pl. xli, f. 18, 18 *a, b, c,* p. 106. *Forest Marble; Laycock.* Walton Collection.

A. olivæformis (Dunker), J. Lycett, Supp. Mon. Moll. Gt. Ool. etc. (1863), pl. xli, f. 4, 4 *a. Forest Marble; Laycock.* Walton Collection.

A. scarburgensis, J. Lycett, Supp. Mon. Moll. Gt. Ool. etc. (1863), pl. xxxi, f. 13, 13 *a,* p. 28. *Cornbrash; Scarborough.* Leckenby Collection.

Another specimen, ventricose variety, W. H. Hudleston, G. M. dec. 3, II (1885), p. 204. *Cornbrash; Scarborough.* Leckenby Collection.

A. suessea, J. Lycett, Supp. Mon. Moll. Gt. Ool. etc. (1863), pl. xlv, f. 29, p. 107. *Forest Marble; Laycock.* Walton Collection.

A. wiltonensis, J. Lycett, Supp. Mon. Moll. Gt. Ool. etc. (1863), pl. xlv, f. 25, p. 107. *Forest Marble; Laycock.* Walton Collection.

Alaria arenosa, W. H. Hudleston, G. M. dec. 3, I (1884), pl. vii, f. 7, p. 198, and Brit. Jurass. Gasterop. Inf. Ool. (1888), pl. iv, f. 1, p. 110. *Dogger Sands; Blue Wyke.* Leckenby Collection.

A. bispinosa (Phillips), W. H. Hudleston, G. M. dec. 2, VII (1880), pl. xvii, f. 6 *c,* p. 533. *Lower Calcareous Grit; Cayton Bay.* Leckenby Collection.

A. bispinosa (Phillips) var. **elegans**, W. H. Hudleston, G. M. dec. 3, I (1884), pl. vi, f. 8, 8 *a,* p. 152. *Cornbrash; Scarborough.* Leckenby Collection.

A. bispinosa (Phillips) var. **pinguis**, W. H. Hudleston, G. M. dec. 3, I (1884), pl. vi, f. 9, 9 *a,* 10, 10 *a,* pp. 152—3. *Kellaways Rock; Scarborough.* Leckenby Collection.

A. hamus (Deslongchamps) var. **phillipsi** (d'Orbigny), W. H.

Hudleston, G. M. dec. 3, I (1884), pl. vi, f. 3, 3 *a*, p. 147 and Brit. Jurass. Gasterop. Inf. Ool. (1888), pl. iv, f. 8 *b*, p. 116. *Dogger Series; Blue Wyke.* Leckenby Collection.

Another specimen, G. M. dec. 3, I (1884), pl. vi, f. 4, p. 149, and Brit. Jurass. Gasterop. Inf. Ool. (1888), pl. iv, f. 8 *c.* p. 116. *Millepore Rock; Stainton Dale.* Leckenby Collection.

Alaria myurus (Deslongchamps), W. H. Hudleston, Brit. Jurass. Gasterop. Inf. Ool. (1888), pl. vi, f. 4 *a*, p. 130. *Inferior Oolite; Dundry.*

A. myurus (Deslongchamps) var. **teres**, W. H. Hudleston, G. M. dec. 3, I (1884), pl. vii, f. 4, 4 *a*, p. 197. *Cornbrash; Scarborough.* Leckenby Collection.

A. pseudo-armata, W. H. Hudleston, G. M. dec. 3, I (1884), pl. vi, f. 6, 6 *a*, p. 150, and Brit. Jurass. Gasterop. Inf. Ool. (1888), pl. v, f. 8, p. 125. *Dogger Series; Blue Wyke.* Leckenby Collection.

A. cf. rarispina, Schlumberger, W. H. Hudleston, Brit. Jurass. Gasterop. Inf. Ool. (1888), pl. iv, f. 12, p. 118. *Inferior Oolite; Bradford Abbas.*

A. sublævigata, W. H. Hudleston, Brit. Jurass. Gasterop. Inf. Ool. (1888), pl. vi, f. 3 *b*, p. 129. *A. myurus?* (Deslongchamps), narrow variety, W. H. Hudleston, G. M. dec. 3, I (1884), pl. vii, f. 6, p. 196. *Millepore Rock; Stainton Dale.* Leckenby Collection.

A. trifida (Phillips), W. H. Hudleston, G. M. dec. 3, I (1884), pl. vi, f. 11, 11 *a*, pl. vii, f. 1, 1 *a*, 3, 3 *a*, pp. 193—5. *Kellaways Rock; Scarborough.* Leckenby Collection.

Another specimen, W. H. Hudleston, G. M. dec. 3, I (1884), pl. vii, f. 2, p. 193. *Oxford Clay; Scarborough.* Leckenby Collection.

A. sp., W. H. Hudleston, G. M. dec. 3, I (1884), pl. vii, f. 5, 5 *a*, p. 198. *Dogger Series; Blue Wyke.* Leckenby Collection.

Amberleya armigera, J. Lycett, Supp. Mon. Moll. Gt. Ool. etc. (1863), pl. xxxi, f. 6, p. 20. Another specimen, W. H. Hudleston, G. M. dec. 3, I (1884), pl. viii, f. 5 *a*, *b*, p. 245. *Cornbrash; Scarborough.* Leckenby Collection.

Amberleya capitanea (Goldfuss), J. Lycett, Supp. Mon. Moll.
Gt. Ool. etc. (1863), pl. xli, f. 1, p. 95. *Forest Marble; Laycock.*
Walton Collection.

A. monilifera, J. Lycett, Supp. Mon. Moll. Gt. Ool. etc. (1863),
pl. xli, f. 10, p. 95. *Forest Marble; Laycock.* Walton Collection.

A. tricincta, J. Lycett, Supp. Mon. Moll. Gt. Ool. etc. (1863),
pl. xli, f. 1, p. 96. *Forest Marble; Laycock.* Walton Collection.

Anisomyon vectis, J. S. Gardner, Q. J. G. S. xxxiii (1877),
pl. ix, f. 3, 4, 5, p. 195. *Lower Greensand; Atherfield.*

Aporrhais globulata (Seeley), J. S. Gardner, G. M. dec. 2, ii
(1875), pp. 54, 295. *Pteroceras globulatum,* H. G. Seeley,
A. M. N. H. ser. 3, vii (1861), pl. xi, f. 1, p. 281. *Tessarolax
globulata,* J. S. Gardner, G. M. dec. 2, vii (1880), p. 53.
Cambridge Greensand; Cambridge.

A. marginata (Sowerby), A. J. Jukes-Browne, Q. J. G. S. xxxiii
(1877), p. 494. *Pterodonta marginata,* H. G. Seeley, A. M.
N. H. ser. 3, vii (1861), pl. xi, f. 2, p. 282. *Cambridge
Greensand; Cambridge.* Presented by J. Carter, Esq.

Avellana ventricosa (Seeley). *Cinulia ventricosa,* H. G. Seeley,
A. M. N. H. ser. 3, vii (1861), p. 293. *Cambridge Greensand;
Cambridge.*

Bellerophon globatus, v. B. subglobatus.

Bellerophon llanvirnensis, H. Hicks, Q. J. G. S. xxxi (1875),
pl. xi, f. 1, p. 188. *Llanvirn Beds; Llanvirn Quarry.* Presented by Dr H. Hicks.

B. ramseyensis, H. Hicks, Q. J. G. S. xxix (1873), pl. iii, f. 30—
32, p. 50. *Tremadoc Beds; Ramsey Island and Tremaenhir.*

B. ruthveni, Salter, Cat. (1873), p. 186. *Upper Ludlow; Benson
Knot, Kendal.*

B. solvensis, H. Hicks, Q. J. G. S. xxix (1873), pl. iii, f. 33,
p. 50. *Tremadoc Beds; Tremaenhir, near Solva.*

B. subdecussatus, F. McCoy, A. M. N. H. ser. 2, vii (1851),
p. 47, and Contrib. Brit. Pal. (1854), p. 182, and Brit. Palæoz.

Foss. (1852), pl. 1 L, f. 25, 25 *a*, p. 311. *Denbighshire Flags;* *Llanwrst, Denbighshire.*

Bellerophon subglobatus, F. McCoy, Brit. Palæoz. Foss. (1852), p. 400. *B. globatus*, J. de C. Sowerby, Trans. Geol. Soc. ser. 2, v (1840), pl. liii, f. 30. *Upper Devonian; Marwood.*

B. sp., Salter, Cat. (1873), p. 173. *Lower Ludlow; Dudley.* Fletcher Collection.

Brachystoma angulare (Seeley), J. S. Gardner, G. M. dec. 2, iv (1877), pl. xvi, f. 22, p. 557. *Scalaria angularis*, H. G. Seeley, A. M. N. H. ser. 3, vii (1861), pl. xi, f. 9, p. 286, (presented by J. Carter, Esq.). A. J. Jukes-Browne, Q. J. G. S. xxxiii (1877), p. 496. *Cambridge Greensand; Cambridge.*

Bulla undulata, Bean, W. H. Hudleston, G. M. dec. 3, ii (1885), pl. v, f. 10, 10 *a*, p. 254. *Cornbrash; Scarborough.* Leckenby Collection.

Calyptræa cooksoniæ (Seeley), J. S. Gardner, Q. J. G. S. xxxiii (1877), p. 201, and G. M. dec. 2, iv (1877), pl. xvi, f. 19, p. 556. *Crepidula cooksoniæ*, H. G. Seeley, A. M. N. H. ser. 3, vii (1861), pl. xi, f. 18, p. 291. A. J. Jukes-Browne, Q. J. G. S. xxxi (1875), p. 294. *Cambridge Greensand; Cambridge.*

Capulus columbinus (Whidborne), G. F. Whidborne, Devonian Fauna S. Eng. (1891), pl. xxi, f. 5, 5 *a*, *b*, p. 214. *Acroculia columbina*, G. F. Whidborne, G. M. dec. 3, vi (1889), p. 30. *Devonian; Lummaton.* Presented by the Rev. G. F. Whidborne, M.A.

C. compressus (Goldfuss), G. F. Whidborne, Devonian Fauna S. Eng. (1891), pl. xx, f. 9, 9 *a*, 10, 10 *a*, 11, 11 *a*, p. 208. *Acroculia prœva*, Eichwald, G. F. Whidborne, G. M. dec. 3, vi (1889), p. 30. *Devonian; Lummaton.* Figs. 10 and 11, presented by the Rev. G. F. Whidborne, M.A.

C. contortus (Römer?), G. F. Whidborne, Devonian Fauna S. Eng. (1891), pl. xxii, f. 7, 7 *a*, 9, 9 *a*, p. 218. *Devonian; Lummaton.* Fig. 7 presented by the Rev. G. F. Whidborne, M.A.

C.? euomphaloides, v. Acroculia euomphaloides.

C. galeritus, G. F. Whidborne, Devonian Fauna S. Eng. (1891), pl. xxii, f. 4, p. 217. *Devonian; Lummaton.* Presented by the Rev. G. F. Whidborne, M.A.

Capulus? invictus, G. F. Whidborne, Devonian Fauna S. Eng. (1891), pl. xix, f. 13, 13 *a*, p. 204. *Devonian; Lummaton.* Presented by the Rev. G. F. Whidborne, M.A.

C. multiplicatus, Giebel, G. F. Whidborne, Devonian Fauna S. Eng. (1891), pl. xxii, f. 11, 11 *a*, p. 220. *Acroculia multiplicata*, G. F. Whidborne, G. M. dec. 3, VI (1889), p. 30. *Devonian; Lummaton.* Presented by the Rev. G. F. Whidborne, M.A.

C. pericompsus, G. F. Whidborne, Devonian Fauna S. Eng. (1891), pl. xx, f. 1—4, p. 205. *Devonian; Lummaton.* Figs. 1, 2, and 4, presented by the Rev. G. F. Whidborne, M.A.

C. puellaris, G. F. Whidborne, Devonian Fauna S. Eng. (1891), pl. xx, f. 12, 12 *a*, 14, 14 *a*, 15, 15 *a*, p. 210. *Devonian; Lummaton.* Presented by the Rev. G. F. Whidborne, M.A.

C. rostratus?, Trenkner, G. F. Whidborne, Devonian Fauna S. Eng. (1891), pl. xx, f. 6, 6 *a*, 7, 7 *a*, *b*, p. 207. *Devonian; Lummaton.* Presented by the Rev. G. F. Whidborne, M.A.

C. squamosus?, Trenkner, G. F. Whidborne, Devonian Fauna S. Eng. (1891), pl. xxi, f. 6, 6 *a*, *b*, p. 215. *Devonian; Lummaton.* Presented by the Rev. G. F. Whidborne, M.A.

C. terminalis, G. F. Whidborne, Devonian Fauna S. Eng. (1891), pl. xx, f. 16, 16 *a*, p. 211. *Devonian; Lummaton.* Presented by the Rev. G. F. Whidborne, M.A.

C. tylotus, G. F. Whidborne, Devonian Fauna S. Eng. (1891), pl. xxi, f. 9, 9 *a*, 10, 10 *a*, pl. xxii, f. 1, 1 *a*, 2, 2 *a*, *b*, p. 216. *Devonian; Lummaton.* Figs. 9, 10, and 2, presented by the Rev. G. F. Whidborne, M.A.

Cassidaria bicatenata (Sowerby). *Cassis bicatenatus*, J. Sowerby, Min. Conch. II (1818), pl. cli, f. 1, 2, p. 117. *Red Crag; Bawdsey, Suffolk.*

Cassis bicatenatus, v. Cassidaria bicatenata.

Cemoria (Puncturella) noachina (Linnæus), S. V. Wood, Crag Mollusca I (1848), pl. xviii, f. 5 *a*, *b*, *c*, p. 166. *Bridlington Crag; Bridlington.* Leckenby Collection.

Cerithinella ? **sp.**, W. H. Hudleston, Brit. Jurass. Gasterop. Inf. Ool. (1890), pl. xii, f. *x*. *Nerinœa ? cingenda* (Sowerby), W. H. Hudleston, G. M. dec. 3, I (1884), pl. iv, f. 9, 9 *a*, p. 112. *Dogger Series; Blue Wyke.* Leckenby Collection.

Cerithium abbreviatum, J. Leckenby, Q. J. G. S. xv (1859), pl. iii, f. 12, p. 13. *Kellaways Rock; the Castle Rock, Scarborough.* Leckenby Collection.

C. bicinctum, W. H. Hudleston, G. M. dec. 2, vii (1880), pl. xvi, f. 4 *a*, *b*, p. 483. *Coral Rag; Malton (probably Langton Wold).* Leckenby Collection.

C.? caninum, v. Fibula canina.

C. circe, d'Orbigny, W. H. Hudleston, Brit. Jurass. Gasterop. Inf. Ool. (1889), pl. x, f. 4, p. 168. *Inferior Oolite; Bridport.*

C. comma, Münster, W. H. Hudleston, Brit. Jurass. Gasterop. Inf. Ool. (1889), pl. x, f. 2, p. 167. *Inferior Oolite (Cadomensis-bed); Oborne, near Sherborne.*

C. comma, Münster, **var.** near to C. unitorquatum, Hébert and Deslongchamps, W. H. Hudleston, Brit. Jurass. Gasterop. Inf. Ool. (1889), pl. x, f. 3, p. 168. *Inferior Oolite (Parkinsoni-zone); Bridport Harbour.*

C. (Kilvertia) comptonese, v. Exelissa weldonis.

C. costigerum, Piette, J. Lycett, Supp. Mon. Moll. Gt. Ool. etc. (1863), pl. xli, f. 11, 11 *a*, *b*, p. 92. *Forest Marble; Laycock.* Walton Collection.

C. culleni, v. Cerithium muricatum *var.* culleni.

C. cf. grandineum, Buvignier, W. H. Hudleston, G. M. dec. 2, vii (1880), pl. xvi, f. 2 *a*, *b*, *c*, p. 482. *Coral Rag; Ayton, near Scarborough.* Leckenby Collection.

C.? hemicinctum, J. Lycett, Supp. Mon. Moll. Gt. Ool. etc. (1863), pl. xli, f. 17, p. 91. *Forest Marble; Laycock.* Walton Collection.

C. leckenbyi, W. H. Hudleston, G. M. dec. 3, I (1884), pl. iii, f. 12, p. 61, and Brit. Jurass. Gasterop. Inf. Ool. (1889), pl. ix, f. 4, p. 158. *Dogger Series; Blue Wyke.* Leckenby Collection.

C. marollinum, d'Orbigny, W. Keeping, Foss. Upware and Brickhill (1883), pl. iii, f. 6, p. 93. *Lower Greensand; Upware.*

7—2

Cerithium? **muricato-costatum**, Münster, W. H. Hudleston, G. M. dec. 2, IX (1882), pl. vi, f. 17, p. 250. *Millepore Bed; Scarborough.* Leckenby Collection.

C. muricatum (Sowerby), W. H. Hudleston, G. M. dec. 3, I (1884), pl. iii, f. 2, p. 53. *Dogger; Blue Wyke.* Leckenby Collection.

C. muricatum (Sowerby) var. **culleni**, Leckenby, W. H. Hudleston, G. M. dec. 3, I (1884), pl. iii, f. 5, p. 54. *C. culleni,* J. Leckenby, Q. J. G. S. XV (1859), pl. iii, f. 13, p. 13. *Kellaways Rock; the Castle Rock, Scarborough.* Leckenby Collection.

C. muricatum (Sowerby) var. **sexlineatum**, W. H. Hudleston, G. M. dec. 3, I (1884), pl. iii, f. 3, 3 *a*, p. 54. *Dogger Sands; Blue Wyke.* Leckenby Collection.

C. muricatum (Sowerby) **var.**, W. H. Hudleston, G. M. dec. 3, I (1884), pl. iii, f. 4, 4 *a*, p. 54. *Cornbrash; Scarborough.* Leckenby Collection.

C. muricatum (Sowerby), extreme varieties, W. H. Hudleston, G. M. dec. 3, I (1884), pl. iii, f. 7, 8, p. 55. *Dogger Sands; Blue Wyke.* Leckenby Collection.

C. tenuistriatum, v. Pyrgiscus tenuistriatus.

C. vetustum (Phillips). *"Chemnitzia" vetusta*, W. H. Hudleston, G. M. dec. 2, IX (1882), pl. vi, f. 9, 11, p. 248. *Fig. 9, Millepore Bed; Cloughton Wyke. Fig. 11, Dogger; Blue Wyke.* Leckenby Collection.

C. vetustum (Phillips) var. **seminudum**, W. H. Hudleston, Brit. Jurass. Gasterop. Inf. Ool. (1889), pl. viii, f. 7, p. 151. *"Chemnitzia" vetusta* var. *seminuda*, W. H. Hudleston, G. M. dec. 2, IX (1882), pl. vi, f. 13, p. 249. *Dogger Series; Blue Wyke.* Leckenby Collection.

C. vetustum (Phillips) **var.**, W. H. Hudleston, Brit. Jur. Gasterop. Inf. Ool. (1889), pl. viii, f. 5 *d. "Chemnitzia" vetusta* var.? *scalariformis*, (Deslongchamps), W. H. Hudleston, G. M. dec. 2, IX (1882), pl. vi, f. 14, p. 249. *Millepore Bed; Cloughton Wyke.* Leckenby Collection. There is some doubt as to which of several specimens is the figured one.

GASTEROPODA. 101

Cerithium vetustum-majus, W. H. Hudleston, Brit. Jurass. Gasterop. Inf. Ool. (1889), pl. viii, f. 6 *b*, p. 150. " *Chemnitzia* " *vetusta major*, W. H. Hudleston, G. M. dec. 2, IX (1882), pl. vi, f. 12, p. 248. *Dogger Series; Blue Wyke.* Leckenby Collection.

C. ? **waltoni**, J. Lycett, Supp. Mon. Moll. Gt. Ool. etc. (1863), pl. xli, f. 16, p. 92. *Forest Marble; Laycock.* Walton Collection.

C. or **Turritella sp.**, W. H. Hudleston, G. M. dec. 3, I (1884), pl. iii, f. 15, 15 *a*, p. 62. *Dogger Sands; Blue Wyke.* Leckenby Collection.

Chemnitzia cf. corallina, d'Orbigny, W. H. Hudleston, G. M. dec. 2, VII (1880), pl. xiii, f. 4, p. 396. *Coral Rag; Malton district (probably Langton Wold).* Leckenby Collection.

C. ferruginea, J. F. Blake and W. H. Hudleston, Q. J. G. S. XXXIII (1877), pl. xiii, f. 5, 5 *a*, p. 393. *Abbotsbury Ironstone (Corallian); Abbotsbury.* Fig. 5 presented by J. F. Walker, Esq., M.A.

C. lineata (Sowerby), W. H. Hudleston, G. M. dec. 2, IX (1882), pl. vi, f. 1 *a, b*, p. 241. *Dogger Series; Blue Wyke.* Leckenby Collection.

Another specimen, J. Leckenby, Q. J. G. S. xv (1859), pl. iii, f. 14, p. 13. W. H. Hudleston, G. M. dec. 2, IX (1882), p. 246. *Kellaways Rock; Scarborough.* Leckenby Collection.

C. lineata-procera, var. **scarburgensis**, Morris and Lycett, W. H. Hudleston, G. M. dec. 2, IX (1882), pl. vi, f. 4, p. 243. *Scarborough Limest ne; Scarborough.* Leckenby Collection.

C. vetusta, v. Cerithium vetustum.

C. vittata (Phillips), W. H. Hudleston, G. M. dec. 2, IX (1882), pl. vi, f. 5 *a, b,* 6, p. 244. *Cornbrash; Scarborough.* Leckenby Collection.

Cinulia ventricosa, v. Avellana ventricosa.

Cloughtonia cincta (Phillips), W. H. Hudleston, G. M. dec. 2, IX (1882), pl. v, f. 14, p. 203. *Scarborough Limestone; Cloughton Wyke, Scarborough.* Leckenby Collection.

Crepidula alta (Seeley), J. S. Gardner, Q. J. G. S. XXXIII (1877), p. 202, and G. M. dec. 2, IV (1877), pl. xvi, f. 21, p. 557. *Cre-*

pidula gaultina?, Buvignier, A. J. Jukes-Browne, Q. J. G. S. XXXI (1875), p. 294. *Galericulus altus*, H. G. Seeley, A. M. N. H. ser. 3, VII (1861), pl. xi, f. 19, p. 292. *Cambridge Greensand; Cambridge.*

Crepidula cooksoniæ, v. Calyptræa cooksoniæ.

Crepidula gaultina, Buvignier, A. J. Jukes-Browne, Q. J. G. S. XXXI (1875), pl. xiv, f. 12, 13 *a*, p. 294. J. S. Gardner, G. M. dec. 2, IV (1877), pl. xvi, f. 20, p. 556. *Cambridge Greensand; Cambridge.*

Cryptaulax cf. undulata (Quenstedt), W. H. Hudleston, Brit. Jurass. Gasterop. Inf. Ool. (1889), pl. xi, f. 14, p. 184. *Inferior Oolite (Parkinsoni-zone); Bridport Harbour.*

Cyclonema crebristria (McCoy). *Turbo crebristria*, F. McCoy, A. M. N. H. ser. 2, VII (1851), p. 49, and Contrib. Brit. Pal. (1854), p. 184, and Brit. Palæoz. Foss. (1852), pl. i K, f. 36, pl. i L, f. 22, p. 295. *Middle Bala; Allt-yr-Anker, Meifod, and Gelli-grin, Bala.*

C. lyrata, v. Trochonema lyrata.

C. undifera (McCoy). *Littorina undifera*, F. McCoy, A. M. N. H. ser. 2, VII (1851), p. 48, and Contrib. Brit. Pal. (1854), p. 183, and Brit. Palæoz. Foss. (1852), pl. i K, f. 46, 46 *a*, p. 306. *Aymestry Limestone; Mortimer's Cross, Aymestry.*

Deslongchampsia eugenei, F. McCoy, in J. Morris and J. Lycett, Moll. Gt. Ool. part i (1850), p. 94. *Great Oolite; Minchin-hampton.*

Eccyliomphalus scoticus, F. McCoy, A. M. N. H. ser. 2, X (1852), p. 195, and Brit. Palæoz. Foss. (1854), p. 246, and Brit. Palæoz. Foss. (1852), pl. i L, f. 15, 15 *a, b*, p. 301. *Lower Bala; Knockdolian.*

Emarginula neocomiensis, d'Orbigny, J. S. Gardner, Q. J. G. S. XXXIII (1877), pl. viii, f. 1, 2, p. 196. *Lower Greensand; Atherfield, Isle of Wight.* Leckenby Collection.

E. unicostata [Seeley MS.], J. S. Gardner, Q. J. G. S. XXXIII (1877), pl. viii, f. 25, 26, p. 199. *Upper Chalk; Norwich.*

Eulima lævigata, Morris and Lycett, W. H. Hudleston, G. M. dec. 2, IX (1882), pl. vi, f. 8, p. 246. Another specimen (missing), J. Lycett, Supp. Mon. Moll. Gt. Ool. etc. (1863), pl. xxxi, f. 3, p. 13. *Cornbrash; Scarborough.* Leckenby Collection.

Euomphalus calyptræa, Salter, Cat. (1873), p. 157. *Wenlock Limestone; Dudley.* Fletcher Collection.

E. costellatus (McCoy). *Straparollus costellatus,* F. McCoy, A. M. N. H. ser. 2, xii (1853), p. 194, and Contrib. Brit. Pal. (1854), p. 262, and Brit. Palæoz. Foss. (1855), pl. iii h, f. 3, 3 *a,* 3 *b,* p. 538. *Carboniferous Limestone; Lowick.*

E. dionysii (De Montfort), G. F. Whidborne, Devonian Fauna S. Eng. (1891), pl. xxiii, f. 20, 20 *a,* p. 244. *Devonian; Lummaton.*

E. lyratus, v. Trochonema lyrata.

E. mariæ, Salter, Cat. (1873), p. 156. *Wenlock Limestone; Dudley.* Fletcher Collection.

E. pacificatus, Salter, Cat. (1873), p. 156. *Wenlock Limestone; Dudley.* Fletcher Collection.

E. triporcatus, v. Trochonema triporcata.

Exelissa weldonis, W. H. Hudleston, Brit. Jurass. Gasterop. Inf. Ool. (1889), pl. xi, f. 8 *a,* p. 179. *Cerithium (Kilvertia) comptonense,* W. H. Hudleston, G. M. dec. 3, i (1884), pl. iii, f. 14, p. 62. *Millepore Rock; Yorkshire Coast.* Leckenby Collection.

Fibula canina, W. H. Hudleston, Brit. Jurass. Gasterop. Inf. Ool. (1889), pl. xi, f. 2 *b,* p. 175. *Cerithium?* *caninum,* W. H. Hudleston, G. M. dec. 3, i (1884), pl. iv, f. 2, p. 108. *Dogger Series; Blue Wyke.* Leckenby Collection.

Funis? brevis, H. G. Seeley, A. M. N. H. ser. 3, vii (1861), pl. xi, f. 8, p. 286. *Cambridge Greensand; Cambridge.*

Fusus quinquecostatus, H. G. Seeley, A. M. N. H. ser. 3, vii (1861), pl. xi, f. 5, p. 284. *Cambridge Greensand; Cambridge.*

F. smithi (Sowerby), A. J. Jukes-Browne, Q. J. G. S. xxxi (1875), pl. xiv, f. 4, 5, p. 291. *Pyrula (Myristica) sowerbyi,* H. G. Seeley, A. M. N. H. ser. 3, vii (1861), p. 283. *Cambridge Greensand; Cambridge.*

F. tricostatus, H. G. Seeley, A. M. N. H. ser. 3, vii (1861), pl. xi,

f. 4, p. 283. *Cambridge Greensand; Cambridge.* Presented by J. Carter, Esq.

Galericulus altus, v. Crepidula alta.

Holopea striatella (Sowerby). *Trochus constrictus,* F. McCoy, A. M. N. H. ser. 2, VII (1851), p. 50, and Contrib. Brit. Pal. (1854), p. 185, and Brit. Palæoz. Foss. (1852), pl. i K, f. 41, p. 296. *Middle Bala; Bryn-melyn quarry, near Bala.*

Holopella duplisulcata, G. F. Whidborne, G. M. dec. 3, VI (1889), p. 30, and Devonian Fauna S. Eng. (1891), pl. xviii, f. 12, 12 *a, b,* p. 227. *Devonian; Lummaton.* Presented by the Rev. G. F. Whidborne, M.A.

H. gracilior, F. McCoy, A. M. N. H. ser. 2, VII (1851), p. 47, and Contrib. Brit. Pal. (1854), p. 182, and Brit. Palæoz. Foss. (1852), pl. i K, f. 33, p. 303. *? Ludlow Beds; Dinas Bran, Llangollen.*

H. intermedia, F. McCoy, A. M. N. H. ser. 2, VII (1851), p. 47, and Contrib. Brit. Pal. (1854), p. 182, and Brit. Palæoz. Foss. (1852), pl. i L, f. 16, p. 304. *Lower Ludlow; High Thorns, Underbarrow.*

H. monile, F. McCoy, A. M. N. H. ser. 2, VII (1851), p. 48, and Contrib. Brit. Pal. (1854), p. 183, and Brit. Palæoz. Foss. (1852), pl. i K, f. 32, 32 *a,* p. 304. *Middle Bala; Selattyn Road, near Chirk.*

H. tenuicincta, F. McCoy, A. M. N. H. ser. 2, VIII (1851), p. 408, and Contrib. Brit. Pal. (1854), p. 230, and Brit. Palæoz. Foss. (1852), pl. i L, f. 17, 17 *a,* p. 304. *May Hill Beds; Mullock, Girvan.*

H. tenuisulcata, Sandberger, G. F. Whidborne, Devonian Fauna S. Eng. (1891), pl. xvii, f. 20, 20 *a,* p. 225. *Devonian; Lummaton.*

Kilvertia pulchra, J. Lycett, Supp. Mon. Moll. Gt. Ool. etc. (1863), pl. xli, f. 12, 12 *a,* p. 94. *Forest Marble; Laycock.* Walton Collection.

Littorina (Trochus) biserta, Phillips, W. H. Hudleston, G. M. dec. 3, I (1884), pl. viii, f. 9 *a, b, c,* p. 248. *Dogger Series; Blue Wyke.* Leckenby Collection.

Littorina cantabrigiensis, W. Keeping, Foss. Upware and Brickhill (1883), pl. iii, f. 11 *a, b,* p. 96. *Lower Greensand; Upware.*

L.? crebricostata [Seeley MS.], A. J. Jukes-Browne, Q. J. G. S. xxxi (1875), pl. xiv, f. 10, 11, p. 292. *Cambridge Greensand; Cambridge.* This specimen cannot be identified with certainty.

L. cf. meriana (Goldfuss), W. H. Hudleston, G. M. dec. 2, vii (1880), pl. xvii, f. 7 *a, b,* p. 534. *Lower Calcareous Grit; Scarborough.* Leckenby Collection. The specimen does not agree exactly with the figure.

L. (Turbo) phillipsi, Morris and Lycett, W. H. Hudleston, G. M. dec. 3, i (1884), pl. viii, f. 1 *a, b,* p. 242. *Millepore Bed; Scarborough.* Leckenby Collection.

L. cf. phillipsi (Morris and Lycett), W. H. Hudleston, G. M. dec. 3, i (1884), pl. viii, f. 3, 4, p. 244. *Fig. 3, Kellaways Rock; Scarborough. Fig. 4, Oxford Clay; Scarborough.* Leckenby Collection.

L. undifera, v. Cyclonema undifera.

L. unicarinata [Bean MS.], W. H. Hudleston, G. M. dec. 3, i (1884), pl. viii, f. 10 *a, b, c,* p. 250. *Dogger Series; Blue Wyke.* Leckenby Collection.

L. upwarensis, W. Keeping, Foss. Upware and Brickhill (1883), pl. iii, f. 10 *a, b,* p. 95. *Lower Greensand; Upware.*

L. ussheri, G. F. Whidborne, Devonian Fauna S. Eng. (1891), pl. xix, f. 6, 6 *a,* p. 188. *Devonian; Lummaton.* Presented by the Rev. G. F. Whidborne, M.A.

L. sp., W. Keeping, Foss. Upware and Brickhill (1883), p. 95. *Lower Greensand; Upware.*

Loxonema elegans, F. McCoy, A. M. N. H. ser. 2, vii (1851), p. 48, and Contrib. Brit. Pal. (1854), p. 183, and Brit. Palæoz. Foss. (1852), pl. i k, f. 34, p. 302. *Lower Ludlow; Green Quarry, Leintwardine.*

L. nexile (Sowerby), F. McCoy, Brit. Palæoz. Foss. (1852), p. 399. *Terebra nexilis,* J. de C. Sowerby, Trans. Geol. Soc. ser. 2, v (1840), pl. liv, f. 17 (larger figure only). *Devonian; South Petherwin.*

Loxonema rœmeri, Kayser, G. F. Whidborne, Devonian Fauna S. Eng. (1891), pl. xvii, f. 19, p. 172. *Devonian; Lummaton.* Presented by the Rev. G. F. Whidborne, M.A.

Maclurea macromphala, v. Ophileta macromphala.

Maclurea magna, Leseur, F. McCoy, Brit. Palæoz. Foss. (1852), pl. i L, f. 13, 14, p. 300. *Lower Bala; Aldons, Girvan.*

Macrochilina brevispirata (McCoy). *Macrochilus brevispiratus*, F. McCoy, A. M. N. H. ser. 2, XII (1853), p. 193, and Contrib. Brit. Pal. (1854), p. 262, and Brit. Palæoz. Foss. (1855), pl. iii H, f. 7, 8, p. 547. *Carboniferous Limestone; fig. 7 from Lowick, fig. 8 from Derbyshire.*

M. elevata, G. F. Whidborne, Devonian Fauna S. Eng. (1891), pl. xvii, f. 11, 11 *a*, p. 170. *Devonian; Lummaton.* Presented by the Rev. G. F. Whidborne, M.A.

M. limnæiforme (McCoy). *Macrochilus limœiformis*, F. McCoy, A. M. N. H. ser. 2, XII (1853), p. 193, and Contrib. Brit. Pal. (1854), p. 261, and Brit. Palæoz. Foss. (1855), pl. iii I, f. 40, p. 548. *Carboniferous Limestone; Lowick.*

M. pupa (Salter). *Macrocheilus pupa*, Salter, Cat. (1873), p. 156. *Wenlock Limestone; Dudley.* Fletcher Collection.

M.? spirata (McCoy). *Macrochilus? spiratus* (McCoy), F. McCoy, Brit. Palæoz. Foss. (1855), pl. iii H, f. 1, 2, p. 549. *Carboniferous Limestone; Lowick.*

M. subcostata (Schlotheim), G. F. Whidborne, Devonian Fauna S. Eng. (1891), pl. xvi, f. 2, 2 *a*, p. 159. *Devonian; Lummaton.* Presented by the Rev. G. F. Whidborne, M.A.

M. subimbricata (d'Orbigny), G. F. Whidborne, Devonian Fauna S. Eng. (1891), pl. xvii, f. 7, p. 166. *Macrocheilus tumescens*, G. F. Whidborne, G. M. dec. 3, VI (1889), p. 30. *Devonian; Lummaton.* Presented by the Rev. G. F. Whidborne, M.A.

Macrochilus, v. Macrochilina.

Mesochilotoma striata, H. G. Seeley, A. M. N. H. ser. 3, VII (1861), p. 284. A. J. Jukes-Browne, Q. J. G. S. XXXIII (1877), p. 496. *Cambridge Greensand; Cambridge.* Presented by J. Carter, Esq., F.G.S.

Monodonta arata, J. Lycett, Supp. Mon. Moll. Gt. Ool. etc.

(1863), pl. xlv, f. 19, 20, p. 102. *Forest Marble; Laycock.* Walton Collection.

Monodonta comma, J. Lycett, Supp. Mon. Moll. Gt. Ool. etc. (1863), pl. xlv, f. 24, 24 *a,* p. 101. *Forest Marble; Farleigh.* Walton Collection.

M. tegulata, J. Lycett, Supp. Mon. Moll. Gt. Ool. etc. (1863), pl. xlv, f. 17, 18, p. 102. *Forest Marble; Laycock.* Walton Collection.

M. waltoni, J. Lycett, Supp. Mon. Moll. Gt. Ool. etc. (1863), pl. xlv, f. 31, 31 *a, b,* p. 101. *Forest Marble; Farleigh.* Walton Collection.

Murchisonia angulata (? Phillips), J. Donald, Q. J. G. S. XLIII (1887), pl. xxiv, f. 2, p. 623. *Carboniferous Limestone; Settle.* Burrow Collection.

M. cancellatula, F. McCoy, A. M. N. H. ser. 2, x (1852), p. 191, and Contrib. Brit. Pal. (1854), p. 243, and Brit. Palæoz. Foss. (1852), pl. i L, f. 20, 20 *a,* p. 292. *May Hill Beds; Mullock Quarry, Dalquharran.*

M. cyclonema, Salter, Cat. (1873), p. 155. *Wenlock Limestone; Dudley.* Fletcher Collection.

M. dispar, F. McCoy, A. M. N. H. ser. 2, XII (1853), p. 191, and Contrib. Brit. Pal. (1854), p. 259, and Brit. Palæoz. Foss. (1855), pl. iii I, f. 37, 37 *a,* p. 531. *Carboniferous Limestone; Lowick.*

M. gyrogonia, F. McCoy, A. M. N. H. ser. 2, x (1852), p. 192, and Contrib. Brit. Pal. (1854), p. 243, and Brit. Palæoz. Foss. (1852), pl. i K, f. 43, 43 *a,* p. 293. *Middle Bala; W. of Llanfechan.* In the explanation of plate i K, the figured specimen is said to be from Yspatty Evan.

M. kendalensis, McCoy, J. Donald, Q. J. G. S. XLIII (1887), pl. xxiv, f. 3, p. 624. *M. verneuiliana,* De Koninck, var. *kendalensis,* F. McCoy, Brit. Palæoz. Foss. (1855), pl. iii H, f. 11, 12, p. 532. *Carboniferous Limestone; Kendal.*

M. pulchra, McCoy, **var.,** F. McCoy, Brit. Palæoz. Foss. (1852), p. 294. *Bala Beds; Allt-yr-Anker, Meifod, and north of Tremadoc. Llandovery Beds; Mathrafal, and Mullock quarry.*

M. simplex, F. McCoy, A. M. N. H. ser. 2, x (1852), p. 192, and Contrib. Brit. Pal. (1854), p. 244, and Brit. Palæoz. Foss. (1852), pl. i κ, f. 44, pl. i ʟ, f. 21, p. 294. *Fig.* 21, *May Hill Beds;* *Mullock, Dalquharran.* *Fig.* 44, *Middle Bala; Allt-yr-Anker.* According to Salter (Cat. Camb. Sil. Foss. p. 83) the specimen from Mullock is not *M. simplex.*

Murchisonia torquata, F. McCoy, Brit. Palæoz. Foss. (1852), pl. i ʟ, f. 19, 19 *a*, p. 294. *Upper Ludlow; Spital, Kendal.*

M. verneuiliana, De Koninck, J. Donald, Q. J. G. S. XLIII (1887), pl. xxiv, f. 6, 7, p. 626. *Carboniferous Limestone; Settle.* Burrow Collection.

M. verneuiliana var. *kendalensis*, v. Murchisonia kendalensis.

Natica adducta, Phillips, var. **canina**, W. H. Hudleston, G. M. dec. 2, IX (1882), pl. v, f. 7, p. 200. *Dogger Series; Blue Wyke.* Leckenby Collection.

N. (Euspira) alta, J. Lycett, Supp. Mon. Moll. Gt. Ool. etc. (1863), pl. xlv, f. 22, 22 *a*, p. 97. *Forest Marble; Laycock.* Walton Collection.

N. arguta, Phillips, W. H. Hudleston, G. M. dec. 2, VII (1880), pl. ix, f. 5 *a*, *b*, p. 297. *Coral Rag? Slingsby.* Leckenby Collection.

N. calypso, d'Orbigny, var. **tenuis**, W. H. Hudleston, G. M. dec. 2, IX (1882), pl. v, f. 13, p. 202. *Lower Calcareous Grit; Cayton Bay.* Leckenby Collection.

N. levistriata, A. J. Jukes-Browne, Q. J. G. S. XXXIII (1877), pl. xxi, f. 6, p. 498. *Cambridge Greensand; Cambridge.*

N. occlusa, S. V. Wood, Crag Mollusca I (1848), pl. xii, f. 4 *a*, *b*, p. 146. *Bridlington Crag; Bridlington.* Leckenby Collection.

N. parva, Sowerby. *Naticopsis? glaucinoides* (Sowerby), F. McCoy, Brit. Palæoz. Foss. (1852), pl. i κ, f. 35, 35 *a*, p. 302. *Upper Ludlow; Benson Knot, near Kendal.*

N. proxima, W. H. Hudleston, G. M. dec. 2, IX (1882), pl. v, f. 8 *a*, *b*, p. 200. *Dogger; Blue Wyke.* Leckenby Collection.

N. punctulata, J. F. Blake, Q. J. G. S. XXXI (1875), p. 224. *Kimeridge Clay; Market Rasen.*

N. punctura (Bean), W. H. Hudleston, G. M. dec. 2, IX (1882), pl. v, f. 11, 12, p. 201. *Fig.* 11, *Cornbrash; Scarborough.* *Fig.* 12, *Kellaways Rock; Scarborough.* Leckenby Collection.

Natica punctura (Bean) var. **bajocensis**, d'Orbigny, W. H. Hudleston, G. M. dec. 2, IX (1882), pl. v, f. 10 *a*, *b*, p. 201. *Dogger Series; Blue Wyke.* Leckenby Collection.

N. texata, J. Lycett, Supp. Mon. Moll. Gt. Ool. etc. (1863), pl. xlv, f. 30, 30 *a*, p. 96. *Forest Marble; Laycock.* Walton Collection.

Naticopsis? glaucinoides, v. Natica parva.

Nerinæa cingenda (Phillips), W. H. Hudleston, G. M. dec. 3, I (1884), pl. iv, f. 3, 3 *a*, 4, 4 *a*, p. 111, and Brit. Jurass. Gasterop. Inf. Ool. (1890), pl. xiv, f. 13 *a*, *b*, p. 210. *Dogger Series; Blue Wyke.* Leckenby Collection.

N. cingenda? (Phillips), W. H. Hudleston, G. M. dec. 3, I (1884), pl. iv, f. 6, 6 *a*, p. 111. *Dogger Series; Blue Wyke.* Leckenby Collection.

N.? cingenda (Sowerby), v. Cerithinella? sp.

N. fusiformis, d'Orbigny, W. H. Hudleston, G. M. dec. 2, VII (1880), pl. xvi, f. 7 *a*, *b*, p. 486. *Coral Rag; Malton.* Leckenby Collection.

N. pseudovisurgis, W. H. Hudleston, G. M. dec. 2, VII (1880), pl. xvii, f. 1, p. 529. *Coralline Oolite; near Pickering.* Leckenby Collection.

N. rœmeri, Phillipi, W. H. Hudleston, Proc. Geol. Assoc. V (1878), pl. iv, f. 2 *a*, *b*, p. 475, and G. M. dec. 2, VII (1880), pl. xvii, f. 2 *a*, *b*, p. 530. *Coralline Oolite; near Scarborough.* Leckenby Collection.

N. tumida, W. Keeping, Foss. Upware and Brickhill (1883), pl. iii, f. 8, p. 94. Probably *Cerithium*, T. Roberts, Q. J. G. S. XLIII (1887), p. 267. *Lower Greensand; Upware.*

N. sp. (? fasciata, Rœmer), W. H. Hudleston, G. M. dec. 3, I (1884), pl. iv, f. 10, p. 114. *Cornbrash; Scarborough.* Leckenby Collection.

N. sp., W. Keeping, Foss. Upware and Brickhill (1883), pl. iii, f. 7, 7 *a*, p. 94. Probably *Cerithium*, T. Roberts, Q. J. G. S. XLIII (1887), p. 267. *Lower Greensand; Upware.*

Nerita buvignieri, J. Lycett, Supp. Mon. Moll. Gt. Ool. etc.

(1863), pl. xli, f. 7, 7 a. *Forest Marble; Farleigh.* Walton Collection.

Nerita minuta, Sowerby, var. **tumidula** (Phillips), W. H. Hudleston, G. M. dec. 3, I (1884), pl. ix, f. 5 a, b, p. 297. *Dogger Series; Blue Wyke.* Leckenby Collection.

N. nodulosa, A. J. Jukes-Browne, Q. J. G. S. xxxIII (1877), pl. xxi, f. 7, 8, p. 499. *Cambridge Greensand; Cambridge.* Presented by J. Carter, Esq., F.G.S.

N. pseudo-costata, d'Orbigny, W. H. Hudleston, G. M. dec. 3, I (1884), pl. ix, f. 7, 8, 9, p. 298. *Dogger Series; Blue Wyke.* Leckenby Collection.

N. (Neritopsis) scalaris, H. G. Seeley, A. M. N. H. ser. 3, VII (1861), pl. xi, f. 13, p. 288. *Cambridge Greensand; Cambridge.* *Neritopsis archiaci,* v. N. canaliculata.

Neritopsis canaliculata (d'Archiac), W. H. Hudleston, G. M. dec. 3, I (1884), pl. ix, f. 12 a, b, p. 301. *N. archiaci* (d'Orbigny), J. Lycett, Supp. Mon. Moll. Gt. Ool. (1863), pl. xxxi, f. 7, 7 a, p. 21. *Cornbrash; Scarborough.* Leckenby Collection.

N. guerrei, Hébert and Deslongchamps, W. H. Hudleston, G. M. dec. 2, VIII (1881), pl. iii, f. 3 c, p. 50. *Coralline Oolite; near Scarborough.* Leckenby Collection.

N. (? Turbo) lævigata (Phillips) **var. A,** W. H. Hudleston, G. M. dec. 3, II (1885), pl. ii, f. 1, p. 50. *Dogger Series; Blue Wyke.* Leckenby Collection.

N. sp., W. H. Hudleston, G. M. dec. 3, I (1884), pl. ix, f. 13 a, b, p. 302. *Kellaways Rock; Scarborough.* Leckenby Collection.

Onustus burtonensis, J. Lycett, Supp. Mon. Moll. Gt. Ool. etc. (1863), pl. xlv, f. 11, 11 a, p. 103. *Forest Marble; Burton Bradstock.* Walton Collection.

O. ornatissimus (d'Orbigny), W. H. Hudleston, G. M. dec. 3, I (1884), pl. ix, f. 1, p. 293. *Dogger Series; Blue Wyke.* Leckenby Collection.

O. pyramidatus (Phillips), W. H. Hudleston, G. M. dec. 3, I (1884), pl. ix, f. 3 a, b, p. 295. *Dogger Series; Blue Wyke.* Leckenby Collection.

GASTEROPODA. 111

Ophileta macromphala (McCoy). *Maclurea macromphala*, F. McCoy, A. M. N. H. ser. 2, x (1852), p. 194, and Contrib. Brit. Pal. (1854), p. 246, and Brit. Palæoz. Foss. (1852), pl. i L, f. 12, 12 *a, b*, p. 300. *Raphistoma* or *Helicotoma*, Salter, Cat. (1873), p. 70. *Middle Bala; Craig Head, Girvan.*

O. sp., H. Hicks, Q. J. G. S. xxxi (1875), pl. xi, f. 3, p. 188. *Llanvirn Beds; Llanvirn Quarry.* Presented by Dr H. Hicks.

Patella sp., W. Keeping, Foss. Upware and Brickhill (1883), pl. iii, f. 14, p. 97. *Lower Greensand; Upware.*

"Phasianella" striata (Sowerby), W. H. Hudleston, G. M. dec. 2, vii (1880), pl. xiv, f. 1 *b*, p. 396. *Coral Rag; Malton (probably from Ayton).* Leckenby Collection.

Phasianella variata, J. Lycett, Supp. Mon. Moll. Gt. Ool. etc. (1863), pl. xlv, f. 28, 28 *a, b*, p. 104. *Forest Marble; Laycock.* Walton Collection.

Philoxene philosophus (Whidborne), G. F. Whidborne, Devonian Fauna S. Eng. (1891), pl. xxiii, f. 17, p. 238. *Phorus philosophus*, G. F. Whidborne, G. M. dec. 3, vi (1889), p. 30. *Devonian; Lummaton.* Presented by the Rev. G. F. Whidborne, M.A.

Phorus philosophus, v. Philoxene philosophus.

Physa bristovi, Forbes, P. de Loriol, Soc. Phys. et d'Hist. Nat. Genève, xviii (1865), pl. ii, f. 9, 9 *a*, 10, 10 *a*, p. 25. *Purbeck Beds; Osmington and Ridgway.* Fisher Collection.

Pileopsis neocomiensis, J. S. Gardner, Q. J. G. S. xxxiii (1877), pl. vii, f. 1, 2, p. 203. *Lower Greensand (Tealby Series); Benniworth Haven, Donnington.*

Platyostoma sigmoidale (Phillips?), G. F. Whidborne, Devonian Fauna S. Eng. (1891), pl. xix, f. 10, 10 *a*, p. 198. *Devonian; Lummaton.* Presented by the Rev. G. F. Whidborne, M.A.

Platyschisma helicoides, Salter, Cat. (1873), p. 186. *Upper Ludlow; Lesmahagow.*

Pleurotoma dowsoni, S. V. Wood, Crag Mollusca, iii (1872), pl. iii, f. 14, p. 39. *Bridlington Crag; Bridlington.* Leckenby Collection. This specimen cannot be found.

Pleurotoma elegantior, S. V. Wood, Crag Mollusca, III (1872), pl. iii, f. 15, p. 38. *Bridlington Crag; Bridlington.* Leckenby Collection.

P. scalaris?, Möller, S. V. Wood, Crag Mollusca, III (1872), pl. iii, f. 12, p. 39. *Bridlington Crag; Bridlington.* Leckenby Collection. This specimen cannot be found.

Pleurotomaria altavittata?, F. McCoy, Brit. Palæoz. Foss. (1855), pl. iii H, f. 9, 10, p. 524. *Carboniferous Limestone; Lowick.*

P. aspera, v. Pleurotomaria interstrialis.

P. bathonica, J. Lycett, Supp. Mon. Moll. Gt. Ool. etc. (1863), pl. xlv, f. 10, p. 105. *Forest Marble; Box Tunnel, near Bath.* Walton Collection.

P. burtonensis, J. Lycett, Supp. Mon. Moll. Gt. Ool. etc. (1863), pl. xlv, f. 8, 8 *a, b*, p. 105. *Forest Marble; Burton Bradstock.* Walton Collection.

P. crenulata, F. McCoy, A. M. N. H. ser. 2, x (1852), p. 190, and Contrib. Brit. Pal. (1854), p. 242, and Brit. Palæoz. Foss. (1852), pl. i K, f. 45, 45 *a*, p. 291. *Upper Ludlow; Brigsteer, Kendal.*

P. decipiens, F. McCoy, A. M. N. H. ser. 2, XII (1853), p. 191 (vars. *a* and *β*), and Contrib. Brit. Pal. (1854), p. 260, and Brit. Palæoz. Foss. (1855), pl. iii H, f. 13, 14, p. 527. *Carboniferous Limestone; Lowick.*

P. depressa (Phillips), W. H. Hudleston, G. M. dec. 3, II (1885), pl. iv, f. 2, 2 *a*, p. 151. *P. striata* [Bean MS.], J. Leckenby, Q. J. G. S. v (1859), pl. iii, f. 2 *a, b*, p. 12. *Kellaways Rock; Castle Hill, Scarborough.* Leckenby Collection.

P. erosa, F. McCoy, A. M. N. H. ser. 2, XII (1853), p. 192, and Contrib. Brit. Pal. (1854), p. 261, and Brit. Palæoz. Foss. (1855), pl. iii I, f. 38, 39, 39 *a, b*, p. 528. *Carboniferous Limestone; Lowick.*

P. ferruginea, W. Keeping, Foss. Upware and Brickhill (1883), pl. iv, f. 2, p. 99. *Lower Greensand; Potton.*

P. fletcheri, Salter, Cat. (1873), p. 154. *Wenlock Limestone; Dudley.* Fletcher Collection.

P. gigantea?, Sowerby, W. Keeping, Foss. Upware and Brickhill (1883), p. 98. *Lower Greensand; Upware.*

Pleurotomaria granulata, J. Lycett (*non* Sowerby), Supp. Mon. Moll. Gt. etc. Ool. (1863), pl. xxxi, f. 8, 8 *a*, p. 24. Another specimen, W. H. Hudleston, G. M. dec. 3, ɪɪ (1885), pl. iii, f. 8, 8 *a*, *b*, p. 127. *Cornbrash; Scarborough*. Leckenby Collection.

Another specimen, W. H. Hudleston, G. M. dec. 3, ɪɪ (1885), pl. iii, f. 9, p. 128. *Kellaways Rock; Scarborough*. Leckenby Collection.

P. guttata (Phillips), W. H. Hudleston, G. M. dec. 3, ɪɪ (1885), pl. iv, f. 4, p. 153. *Kellaways Rock; Scarborough*. Leckenby Collection.

P. interstrialis, Phillips, F. McCoy, Brit. Palæoz. Foss. (1852), p. 398. *Pleurotomaria aspera*, J. de C. Sowerby, Trans. Geol. Soc. ser. 2, ᴠ (1840), pl. liv, f. 16. *Devonian; South Petherwin*.

P. llanvirnensis, H. Hicks, Q. J. G. S. xxxɪ (1875), pl. xi, f. 4, p. 188. *Llanvirn Beds; Llanvirn Quarry*. Presented by Dr H. Hicks, F.R.S.

P. renevieri, W. Keeping, Foss. Upware and Brickhill (1883), pl. iv, f. 1 *a*, *b*, p. 98. *Lower Greensand; Upware*.

P. reticulata (Sowerby). *Trochus reticulatus*, J. Sowerby, Min. Conch. ɪɪɪ (1821), pl. cclxxii, f. 2, p. 128. *Coral Rag; Ringstead Bay*. (Not from the Kimeridge Clay as stated by Sowerby.)

P. semiconcava, H. G. Seeley, A. M. N. H. ser. 3, ᴠɪɪ (1861), pl. xi, f. 17, p. 291. *Cambridge Greensand; Cambridge*.

P. striata, v. Pleurotomaria depressa.

P. striatissima, Salter, Cat. (1873), p. 171. *Lower Ludlow; Green Quarry, Leintwardine: and Wenlock Shale; Dudley*.

Pseudomelania gracilis, W. H. Hudleston, G. M. dec. 2, ᴠɪɪ (1880), pl. xiv, f. 3 *a*, p. 398. *Coral Rag; near Ayton*. Leckenby Collection.

Pteroceras globulatum, v. Aporrhais globulata.

Pterodonta marginata, v. Aporrhais marginata.

Purpurina elaborata (Lycett), W. H. Hudleston, Brit. Jurass. Gasterop. Inf. Ool. (1888), pl. i, f. 1 *g*, p. 85. *P. elaborata*, var. *bajocensis*, W. H. Hudleston, G. M. dec. 2, ix (1882), pl. v, f. 2, p. 195. *Dogger Series; Blue Wyke.* Leckenby Collection.

Purpurina (Eucycloidea) "**fusiforme**" (species or variety related to *P. bianor*, d'Orbigny), W. H. Hudleston, Brit. Jurass. Gasterop. Inf. Ool. (1888), pl. ii, f. 6 *a, b*, p. 96. *Inferior Oolite; Yeovil.*

Purpuroidea cf. tuberosa (Sowerby), W. H. Hudleston, G. M. dec. 2, vii (1880), pl. viii, f. 3 *a, b*, p. 292. *Coral Rag; Langton Wold.* Leckenby Collection. (Originally in Bean's Collection.)

Pyrgiscus tenuistriatus (Seeley), J. S. Gardner, G. M. dec. 2, vi (1876), p. 112. *Cerithium tenuistriatum*, H. G. Seeley, A. M. N. H. ser. 3, vii (1861), pl. xi, f. 6, p. 285. *Cambridge Greensand; Cambridge.*

Pyrula (Myristica) sowerbyi, v. Fusus smithi.

Rissoina subulata, J. Lycett, Supp. Mon. Moll. Gt. Ool. etc. (1863), pl. xli, f. 9, p. 98. *Forest Marble; Laycock.* Walton Collection. This specimen cannot be identified with certainty.

Scalaria angularis, v. Brachystoma angulare.

Scalaria fasciata, R. Etheridge, Appendix to W. H. Penning and A. J. Jukes-Browne, Geol. Cambridge (1881), pl. i, f. 1, p. 140. *Lower Chalk (Totternhoe Stone); Burwell.*

Solarium carteri, H. G. Seeley, A. M. N. H. ser. 3, vii (1861), pl. xi, f. 12, p. 288. *Cambridge Greensand; Cambridge.*

S. planum, H. G. Seeley, A. M. N. H. ser. 3, vii (1861), pl. xi, f. 11, p. 287. *Cambridge Greensand; Cambridge.* Presented by J. Carter, Esq., F.G.S.

S. sedgwicki, H. G. Seeley, A. M. N. H. ser. 3, vii (1861), pl. xi, f. 10, p. 286. *Cambridge Greensand; Cambridge.*

S. turbiniformis, J. Lycett, Supp. Mon. Moll. Gt. Ool. etc. (1863), pl. xlv, f. 23, 23 *a, b*, p. 104. *Great Oolite; Hampton Cliffs, near Bath.* Walton Collection.

S. waltoni, J. Lycett, Supp. Mon. Moll. Gt. Ool. etc. (1863), pl. xlv, f. 26, 26 *a, b, c*, p. 104. *Great Oolite; below Hampton Cliffs.* Walton Collection.

Stomatodon politus, H. G. Seeley, A. M. N. H. ser. 3, VII (1861), pl. xi, f. 22, p. 293. *Cambridge Greensand; Cambridge.*

Straparollus costellatus, v. Euomphalus costellatus.

Tectura tenuistriata (Seeley), J. S. Gardner, Q. J. G. S. XXXIII (1877), pl. vii, f. 18, p. 194. *Acmœa tenuistriata*, H. G. Seeley, A. M. N. H. ser. 3, VII (1861), pl. xi, f. 20, p. 292. *Cambridge Greensand; Cambridge.*

Terebra nexilis, v. Loxonema nexile.

Tornatella pyrostoma, H. G. Seeley, A. M. N. H. ser. 3, VII (1861), pl. xi, f. 21, p. 292. *Cambridge Greensand; Cambridge.*

Trichotropis borealis, Broderip and Sowerby, S. V. Wood, Crag Mollusca, I (1848), pl. xix, f. 11 *a*, *b*, p. 67. *Bridlington Crag; Bridlington.* Leckenby Collection.

Trochonema bijugosa, Salter, Cat. (1873), p. 156. *Wenlock Limestone; Dudley.* Fletcher Collection.

T. lyrata (McCoy). *Euomphalus lyratus*, F. McCoy, A. M. N. H. ser. 2, X (1852), p. 193, and Contrib. Brit. Pal. (1854), p. 245, and Brit. Palæoz. Foss. (1852), pl. i L, f. 23, p. 298. *Cyclonema lyrata*, Salter, Cat. (1873), p. 69. *Middle Bala; Llansaintffraid, Glyn Ceiriog.*

T. triporcata (McCoy). *Euomphalus triporcatus*, F. McCoy, A. M. N. H. ser. 2, X (1852), p. 193, and Contrib. Brit. Pal. (1854), p. 245, and Brit. Palæoz. Foss. (1852), pl. i K, f. 37 [38 missing], p. 299. *Middle Bala; Cyrn-y-brain, Wrexham.*

Trochotoma calix (Phillips), W. H. Hudleston, G. M. dec. 3, II (1885), p. 157. *Dogger; Blue Wyke.* Leckenby Collection.

Trochus acuticarina, Buvignier, W. H. Hudleston, Proc. Geol. Assoc. V (1878), pl. iv, f. 5, p. 476, and G. M. dec. 2, VIII (1881), pl. iii, f. 10, p. 57. *Coral Rag; Malton.* Leckenby Collection.

T. aytonensis, Blake and Hudleston, W. H. Hudleston, G. M. dec. 2, VIII (1881), pl. iii, f. 12, p. 58. *Coralline Oolite; Malton.* (Pickering, according to Mr Hudleston.) Leckenby Collection.

T. burtonensis, J. Lycett, Supp. Mon. Moll. Gt. Ool. etc. (1863), pl. xlv, f. 16, p. 99. *Forest Marble; Burton Bradstock.* Walton Collection.

Trochus cælatulus, F. McCoy, A. M. N. H. ser. 2, VII (1851), p. 49, and Contrib. Brit. Pal. (1854), p. 184, and Brit. Palæoz. Foss. (1852), pl. i K, f. 40, 40 *a*, p. 296. *Woolhope Limestone; Old Radnor, Presteign.*

T. cancellatus, Seeley, A. J. Jukes-Browne, Q. J. G. S. XXXIII (1877), p. 498. *T. tollotianus*, Pictet and Roux, A. J. Jukes-Browne, Q. J. G. S. XXXI (1875), pl. xiv, f. 7—9, p. 292. Another specimen, *Trochus (Trochodon) cancellatus*, H. G. Seeley, A. M. N. H. ser. 3, VII (1861), pl. xi, f. 15, p. 290. *Cambridge Greensand; Cambridge.*

T. concavus, v. Trochus ziziphinus.

T. constrictus, v. Holopea striatella.

T. cf. dimidiatus, Sowerby, W. H. Hudleston, G. M. dec. 3, II (1885), pl. ii, f. 15, p. 58. *Dogger Sands; Blue Wyke.* Leckenby Collection.

T. granularis, W. H. Hudleston, G. M. dec. 2, VIII (1881), pl. iii, f. 11 *a*, *b*, p. 57. *Coralline Oolite; Malton.* Leckenby Collection.

T.? leckenbyi, J. Morris and J. Lycett, Mollusca Gt. Ool. part i (1850), pl. xv, f. 21, 21 *a*, p. 115. W. H. Hudleston, G. M. dec. 3, II (1885), pl. iii, f. 7, p. 126. *Scarborough Limestone; Cloughton Wyke, near Scarborough.* Leckenby Collection.

T. (Gibbula) levistriatus, H. G. Seeley, A. M. N. H. ser. 3, VII (1861), pl. xi, f. 16, p. 290. *Cambridge Greensand; Cambridge.*

T. monilitectus, Phillips, **var. B**, W. H. Hudleston, G. M. dec. 3, II (1885), p. 122. *Cornbrash; Scarborough.* Leckenby Collection.

T. moorei, F. McCoy, A. M. N. H. ser. 2, VII (1851), p. 50, and Contrib. Brit. Pal. (1854), p. 185, and Brit. Palæoz. Foss. (1852), pl. i L, f. 18, 18 *a*, p. 297. *May Hill Beds; Dalquharran, Ayrshire.*

T. reticulatus, v. Pleurotomaria reticulata.

T. scarburgensis, W. H. Hudleston, G. M. dec. 3, II (1885), pl. iii, f. 2, 2 *a*, p. 123. *Cornbrash; Scarborough.* Leckenby Collection.

Trochus strigosus, J. Lycett, Supp. Mon. Moll. Gt. Ool. etc. (1863), pl. xlv, f. 12, p. 29. W. H. Hudleston, G. M. dec. 3, II (1885), pl. iii, f. 3, 3 *a*, 4, p. 124. *Cornbrash; Scarborough.* Leckenby Collection.

T. subglaber, W. H. Hudleston, G. M. dec. 3, II (1885), pl. iii, f. 6, 6 *a*, p. 125. *Cornbrash; Scarborough.* Leckenby Collection.

T. tollotianus, v. Trochus cancellatus.

T. ziziphinus, Linnæus. *T. concavus*, J. Sowerby, Min. Conch. III (1821), pl. cclxxii, f. 1, p. 127. *Coralline Crag; Orford, Suffolk.*

T. sp. or **var.**, W. H. Hudleston, G. M. dec. 3, II (1885), p. 126 (footnote). *Cornbrash; Scarborough.* Leckenby Collection.

Trophon craticulatus, Fabricius, S. V. Wood, Crag Mollusca III (1872), pl. iii, f. 1 *a, b*, p. 25. *Bridlington Crag; Bridlington.* Leckenby Collection.

T. sabini, Hancock, S. V. Wood, Crag Mollusca, III (1872), pl. ii, f. 15 *c*, p. 23. *Bridlington Crag; Bridlington.* Leckenby Collection.

T. scalariformis, Gould, S. V. Wood, Crag Mollusca, III (1872), pl. iii, f. 10 *a, b*, p. 26. *Bridlington Crag; Bridlington.* Leckenby Collection. This specimen cannot be found.

T. ventricosus? (Gray), S. V. Wood, Crag Mollusca, III (1872), pl. iii, f. 4, p. 22. *Bridlington Crag; Bridlington.* Leckenby Collection.

Turbo burtonensis, J. Lycett, Supp. Mon. Moll. Gt. Ool. etc. (1863), pl. xlv, f. 15, p. 100. *Forest Marble; Burton Bradstock.* Walton Collection. The specimen does not agree with the figure.

T. crebristria, v. Cyclonema crebristria.

T. depauperatus (Morris and Lycett), J. Lycett, Supp. Mon. Moll. Gt. Ool. etc. (1863), pl. xlv, f. 13, p. 99. *Forest Marble; Laycock.* Walton Collection.

T. (Delphinula) funiculatus, Phillips, varieties W. H. Hudleston, G. M. dec. 3, II (1885), pl. ii, f. 7, 8, p. 54. *Fig. 7, Dogger; Blue Wyke. Fig. 8, Cornbrash; Scarborough.* Leckenby Collection.

118   GASTEROPODA.

Turbo (Delphinula) granatus [Bean MS.], W. H. Hudleston, G. M. dec. 3, II (1885), pl. ii, f. 9, 9 *a*, p. 55. *Dogger Series; Blue Wyke.* Leckenby Collection.

T. (Monodonta) lævigatus (Sowerby), **var.**, W. H. Hudleston, G. M. dec. 3, II (1885), pl. ii, f. 6, 6 *a*, p. 54. *Dogger Series; Blue Wyke.* Leckenby Collection.

T. lævis, Buvignier, W. H. Hudleston, Proc. Geol. Assoc. v (1878), pl. iv, f. 6, p. 476. *Coral Rag; Ayton.* Leckenby Collection.

T. pictetianus, d'Orbigny, A. J. Jukes-Browne, Q. J. G. S. XXXIII (1877), pl. xxi, f. 3—5, p. 497. *Turboidea nodosa,* H. G. Seeley, A. M. N. H. ser. 3, VII (1861), pl. xi, f. 14, p. 289. *Cambridge Greensand; Cambridge.* Presented by J. Carter, Esq.

    Another specimen, A. J. Jukes-Browne, Q. J. G. S. XXXIII (1877), pl. xxi, f. 5, p. 497. *Gault; Perte du Rhône.*

T. reedi, W. Keeping, Foss. Upware and Brickhill (1883), pl. iii, f. 13 *a, b,* p. 97. *Lower Greensand; Upware.*

T. subtexatus, J. Lycett, Supp. Mon. Moll. Gt. Ool. etc. (1863), pl. xli, f. 15, 15 *a*, p. 100. *Forest Marble; Farleigh.* Walton Collection.

"**T.**" **sulcostomus**, Phillips, W. H. Hudleston, G. M. dec. 3, I (1884), pl. viii, f. 6, 6 *a*, p. 246. *Kellaways Rock; Scarborough.* Leckenby Collection.

Turboidea nodosa, v. Turbo pictetianus.

Turritella opalina, Quenstedt, var. **canina**, W. H. Hudleston, G. M. dec. 3, I (1884), pl. vii, f. 10, 10 *a*, p. 201. *Dogger Series; Blue Wyke.* Leckenby Collection.

T. quadrivittata, Phillips, W. H. Hudleston, G. M. dec. 3, I (1884), pl. vii, f. 12, p. 202. *Dogger Sands; Blue Wyke.* Leckenby Collection.

T. sp., W. H. Hudleston, G. M. dec. 3, I (1884), pl. vii, f. 13, 13 *a*, p. 203. *Dogger; Blue Wyke.* Leckenby Collection.

Valvata helicoides, Forbes, P. de Loriol, Soc. Phys. et d'Hist. Nat. Genève XVIII (1865), pl. ii, f. 23, 23 *a, b, c,* p. 33. *Purbeck Beds; Ridgway.* Fisher Collection.

Voluta lamberti, J. Sowerby, Min. Conch. II (1818), pl. cxxix, f. 2, p. 65. *Red Crag.* Sowerby gives *Bawdsey* as the locality, but the specimen is labelled *Sutton*.

POLYPLACOPHORA.

Chiton burrowianus, J. W. Kirkby, Q. J. G. S. XVIII (1862), p. 234, woodcuts 1, 2. *Carboniferous Limestone ; Settle.* Burrow Collection.

C. coloratus, J. W. Kirkby, Q. J. G. S. XVIII (1862), p. 234, woodcuts 3—6. J. W. Kirkby and J. Young, G. M. IV (1867), pl. xvi, f. 8 *a, b,* p. 340. *Carboniferous Limestone ; Settle.* Burrow Collection.

C. cordatus?, Kirkby, J. W. Kirkby and J. Young, G. M. IV (1867), pl. xvi, f. 10, 11 *a, b,* p. 341. *Carboniferous Limestone ; Settle.* Burrow Collection.

C. sp., J. W. Kirkby, Q. J. G. S. XVIII (1862), p. 236, woodcuts 9, 10. *Carboniferous Limestone ; Settle.* Burrow Collection.

C. sp., J. W. Kirkby, Q. J. G. S. XVIII (1862), p. 235, woodcuts 7, 8. *Carboniferous Limestone ; Settle.* Burrow Collection.

Chitonellus subantiquus, J. W. Kirkby and J. Young, G. M. IV (1867), pl. xvi, f. 12 *a, b,* 13, p. 341. *Carboniferous Limestone ; Settle.* Burrow Collection.

PTEROPODA.

Conularia bifasciata, Salter, Cat. (1873), p. 171. *Lower Ludlow ; Leintwardine.*

C. clavus, Salter, Cat. (1873), p. 153. *Wenlock Limestone; Dudley.* Fletcher Collection.

C. homfrayi, J. W. Salter, Appendix to A. C. Ramsay, Geol. North Wales (1866), pl. x, f. 11—13, p. 354, and *ibid.* second edition (1881), p. 562. *Upper Tremadoc ; Garth, Portmadoc.* One specimen (fig. 13) presented by D. Homfray, Esq.

120 PTEROPODA.

Conularia llanvirnensis, H. Hicks, Q. J. G. S. XXXI (1875), pl. xi, f. 5, 6, p. 189. *Llanvirn Beds; Llanvirn Quarry.* Presented by Dr H. Hicks, F.R.S.

C. subtilis, J. W. Salter, Appendix to F. McCoy, Brit. Palæoz. Foss. (1855), p. vi. F. McCoy, *ibid.* (1852), pl. i L, f. 24, 24 a, p. 288. *Upper Ludlow; Benson Knot, Kendal.*

Creseis primæva, v. Orthoceras primævum.

C. sedgwicki, v. Orthoceras subundulatum.

Cyrtotheca hamula, H. Hicks, Q. J. G. S. XXVIII (1872), pl. vii, f. 14, p. 179. *Menevian Beds; Porth-y-rhaw, St Davids.*

Graptotheca catenulata, Salter, Cat. (1873), p. 171. *Lower Ludlow; Leintwardine.*

Hyolithes antiquus (Hicks). *Theca antiqua*, H. Hicks, Q. J. G. S. XXVII (1871), pl. xvi, f. 13, p. 400. *Harlech Beds; St Davids.* Presented by Dr H. Hicks.

H. caereesiensis (Hicks). *Theca caereesiensis*, H. Hicks, Q. J. G. S. XXXI (1875), pl. xi, f. 7, p. 189. *Llanvirn Beds; Llanvirn Quarry.* Presented by Dr H. Hicks.

H. corrugatus (Salter). *Theca corrugata*, Salter, H. Hicks, Q. J. G. S. XXVIII (1872), pl. vii, f. 17. *Menevian Beds; Porth-y-rhaw, St Davids.*

H. davidi (Hicks). *Theca davidi*, H. Hicks, Q. J. G. S. XXIX (1873), pl. iii, f. 28, p. 50. *Tremadoc Beds; Ramsey Island.* Presented by Dr H. Hicks.

H. harknessi (Hicks). *Theca harknessi*, H. Hicks, in Salter, Cat. (1873), p. 24. H. Hicks, Q. J. G. S. XXXI (1875), pl. x, f. 11, p. 189. *Middle Arenig; Whitesand Bay.* Presented by Dr H. Hicks.

H. homfrayi (Salter). *Theca homfrayi*, Salter, Cat. (1873), p. 8. *Menevian Beds; Tyddyngwladis, near Dolgelly.* Presented by D. Homfray, Esq.

H. penultimus (Hicks). *Theca penultima*, H. Hicks, Rep. Brit. Assoc. (1865), p. 285, and Q. J. G. S. XXVIII (1872), pl. vii, f. 15, 16, p. 180. *Menevian Beds; Porth-y-rhaw, St Davids.*

Hyolithes stiletto (Hicks). *Theca stiletto*, H. Hicks, Q. J. G. S. XXVIII (1872), pl. vii, f. 18, 19, p. 180. *Menevian Beds; Porth-y-rhaw, St Davids.*

H. sulcatus (Salter). *Theca sulcata*, Salter, Cat. (1873), p. 17. *Upper Tremadoc; Llanerch.* Presented by D. Homfray, Esq.

H. trilineatus (Salter). *Theca trilineata*, Salter, Cat. (1873), p. 18. *Base of Upper Tremadoc; Moel-y-gest.* Presented by D. Homfray, Esq.

Stenotheca cornucopia, Salter, Cat. (1873), p. 8. H. Hicks, Q. J. G. S. XXVIII (1872), pl. vii, f. 12, p. 180. *Menevian Beds; Porth-y-rhaw, St Davids.*

Theca, v. Hyolithes.

SCAPHOPODA.

Dentalium jeffreysi, J. S. Gardner, Q. J. G. S. XXXIV (1878), pl. iii, f. 30, p. 61. *Lower Greensand; Atherfield.*

CEPHALOPODA.

Acanthoteuthis tricarinata, v. Plesioteuthis tricarinata.

Ammonites acanthonotus, H. G. Seeley, A. M. N. H. ser. 3, XVI (1865), pl. xi, f. 5, p. 234. Malformed individual of *A. auritus*, Sowerby, A. J. Jukes-Browne, Q. J. G. S. XXXIII (1877), p. 491. *Cambridge Greensand; Cambridge.*

Ammonites (Aegoceras) acuticostatum, T. Wright, Lias Ammonites (1880), pl. xxxv, f. 1—3, 7, p. 371. *Lias (zone of A. Jamesoni); Robin Hood's Bay.* Leckenby Collection.

A. alligatus [Bean MS.], J. Leckenby, Q. J. G. S. XV (1859), pl. ii, f. 2 a, b, p. 9. *Kellaways Rock; Scarborough Castle.* Leckenby Collection.

A. (Aegoceras) armatus, Sowerby, T. Wright, Lias Ammonites (1880), pl. xxviii, f. 1—4, p. 340. *Lower Lias; Robin Hood's Bay.* Leckenby Collection.

A. (Aegoceras) belcheri, Simpson, T. Wright, Lias Ammonites

(1879), pl. xv, f. 7—9, p. 313. *Lower Lias (zone of A. pla-norbis)*; *Robin Hood's Bay.* Leckenby Collection.

Ammonites biplex, Sowerby, W. Keeping, Foss. Upware and Brickhill (1883), pl. viii, f. 11, p. 155. *Lower Greensand* (derived); *Upware.*

A. (Aegoceras) capricornus, Schlotheim, T. Wright, Lias Ammonites (1880), pl. xxxiv, f. 1—8, p. 368. *Middle Lias; Robin Hood's Bay.* Leckenby Collection.

A. cœlonotus, H. G. Seeley, A. M. N. H. ser. 3, xvi (1865), pl. x, f. 2, p. 237. *Cambridge Greensand; Cambridge.*

A. cœlonotus, Seeley, var. **valbonnensis**, Hébert and Munier-Chalmas, A. J. Jukes-Browne, Q. J. G. S. xxxiii (1877), p. 491. *A. cœlonotus*, var., H. G. Seeley, A. M. N. H. xvi (1865), pl. x, f. 3, p. 237. *Cambridge Greensand; Cambridge.*

A. (Arietites) collenotti, d'Orbigny, T. Wright, Lias Ammonites (1881), pl. vi, f. 1, pl. xxii A, f. 6—9, pl. xxii B, f. 1—3, p. 304. The specimen figured pl. vi, f. 1 cannot be found. *Lower Lias; Robin Hood's Bay, near Whitby.* Leckenby Collection.

A. cornuelianus, d'Orbigny, W. Keeping, Foss. Upware and Brickhill (1883), pl. i, f. 9 *a—c*, p. 89. *Lower Greensand; Upware.*

A. cratus, H. G. Seeley, A. M. N. H. ser. 3, xvi (1865), pl. xi, f. 2, p. 240. *Cambridge Greensand; Cambridge.*

A. (Aegoceras) curvicornis, Schloenbach, T. Wright, Lias Ammonites (1880), pl. xxxi, f. 3, 4, p. 377. *Middle Lias; Charmouth.* Leckenby Collection.

A. (Aegoceras) davœi, Sowerby, T. Wright, Lias Ammonites (1880), pl. xxxi, f. 1, 2, p. 346. *Middle Lias; Golden Cap, near Chideock.* Leckenby Collection.

A. deshayesi, Leymerie, W. Keeping, Foss. Upware and Brickhill (1883), pl. iii, f. 1, p. 90. *Lower Greensand; Upware.*

A. glossonotus, H. G. Seeley, A. M. N. H. ser. 3, xvi (1865), pl. x, f. 4, p. 235. *Cambridge Greensand; Ashwell.*

A. gowerianus, Sowerby, J. Leckenby, Q. J. G. S. xv (1859), pl. i, f. 1 *b*, p. 9. *Kellaways Rock; Red Cliff, near Scarborough.* Leckenby Collection.

Ammonites henslowi, v. Goniatites henslowi.

A. (**Aegoceras**) **heterogenes**, Young and Bird, T. Wright, Lias Ammonites (1880), pl. xxxv, f. 4—6, p. 370. *Middle Lias (zone of A. capricornus); Robin Hood's Bay.* Leckenby Collection.

A. **hyperbolicus** [Simpson MS.], J. Leckenby, Q. J. G. S. xv (1859), pl. ii, f. 4 *a, b,* p. 12. *Kellaways Rock; Red Cliff, Scarborough.* Leckenby Collection.

A. (**Arietites**) **impendens**, Young and Bird, T. Wright, Lias Ammonites (1881), pl. xxii A, f. 1—5, p. 302. *Lias (zone of A. oxynotus); Robin Hood's Bay.* Leckenby Collection.

A. **latidorsatus**, Michelin, H. G. Seeley, A. M. N. H. ser. 3, xvi (1865), p. 229. *Cambridge Greensand; Cambridge.* Presented by the Rev. J. F. Blake, M.A.

A. (**Aegoceras**) **leckenbyi**, T. Wright, Lias Ammonites (1880), pl. xxx, f. 1—7, p. 344. *Lower Lias (zone of A. armatus); Whitby.* Leckenby Collection.

A. **leptus**, H. G. Seeley, A. M. N. H. ser. 3, xvi (1865), pl. x, f. 5, p. 240. *Cambridge Greensand; Ashwell.*

A. (**Harpoceras**) **lythensis**, Young and Bird, T. Wright, Lias Ammonites (1882), pl. lxii, f. 4—6, p. 444. *Upper Lias; Whitby.* Leckenby Collection.

A. (**Olcostephanus**) **multiplicatus**, Roemer, A. Pavlow, Bull. Soc. Imp. Nat. Mosc. (1889), pl. iii, f. 2. *Lower Greensand (Tealby Series); Donnington.*

A. **navicularis**, Mantell, var. **nothus**, H. G. Seeley, A. M. N. H. ser. 3, xvi (1865), p. 228. *Cambridge Greensand; Cambridge.*

A. **ochetonotus**, H. G. Seeley, A. M. N. H. ser. 3, xvii (1866), p. 174. *Red Chalk; Hunstanton.*

A. **pachys**, H. G. Seeley, A. M. N. H. ser. 3, xvi (1865), pl. xi, f. 4, p. 227. *Cambridge Greensand; Cambridge.*

A. **placenta** [Simpson MS.], J. Leckenby, Q. J. G. S. xv (1859), pl. ii, f. 1, p. 10. *Kellaways Rock; the Castle Rock, Scarborough.* Leckenby Collection.

Ammonites (Aegoceras) planorbis, Sowerby, T. Wright, Lias Ammonites (1879), pl. xiv, f. 1, 2, p. 308. *Lower Lias (zone of A. planorbis); Robin Hood's Bay.* Leckenby Collection.

A. planulatus, Sowerby, var. **mayorianus**, d'Orbigny, H. G. Seeley, A. M. N. H. ser. 3, xvi (1865), p. 229. *Cambridge Greensand; Cambridge.*

A. poculum [Bean MS.], J. Leckenby, Q. J. G. S. xv (1859), pl. i, f. 4 *a, b, c,* p. 9. *Kellaways Rock; near Gristhorpe Bay.* Leckenby Collection.

A. putealis [Bean MS.], J. Leckenby, Q. J. G. S. xv (1859), pl. ii, f. 3 *a, b, c,* p. 11. *Kellaways Rock; the Castle Rock, Scarborough.* Leckenby Collection.

A. renauxianus, d'Orbigny, **var.**, H. G. Seeley, A. M. N. H. ser. 3, xvi (1865), pl. xi, f. 6, p. 243. *Cambridge Greensand; Cambridge.*

A. reversus [Simpson MS.], J. Leckenby, Q. J. G. S. xv (1859), pl. i, f. 2, p. 9. *Kellaways Rock; the Castle Rock, Scarborough.* Leckenby Collection.

A. rhamnonotus, H. G. Seeley, A. M. N. H. ser. 3, xvi (1865), pl. xi, f. 7, p. 233. A. J. Jukes-Browne, Q. J. G. S. xxxiii (1877), p. 490. *Cambridge Greensand; Cambridge.*

A. rugosus, J. Leckenby, Q. J. G. S. xv (1859), p. 9. *Kellaways Rock; near Gristhorpe Bay.* Leckenby Collection.

A. sexangulatus, H. G. Seeley, A. M. N. H. ser. 3, xvi (1865), pl. xi, f. 1, p. 233. *Cambridge Greensand; Cambridge.*

A. sphærotus, H. G. Seeley, A. M. N. H. ser. 3, xvii (1866), p. 175. *Red Chalk; Hunstanton.*

A. (Amaltheus) simpsoni [Bean MS.], T. Wright, Lias Ammonites (1881), pl. xlvii, f. 4—7, p. 392. *Lias (zone of A. oxynotus); Robin Hood's Bay.* Leckenby Collection.

A. (Olcostephanus) stenomphalus, A. Pavlow, Bull. Soc. Imp. Nat. Mosc. (1889), pl. iii, f. 1, p. 59. *Lower Greensand; Donnington.*

A. timotheanus, Pictet, H. G. Seeley, A. M. N. H. ser. 3, xvi (1865), p. 228. *Cambridge Greensand; Cambridge.* Presented by J. Carter, Esq., F.G.S.

Ammonites vertumnus [Bean MS.], J. Leckenby, Q. J. G. S. xv (1859), pl. i, f. 3 *a*, *b*, p. 9. *Kellaways Rock; near Gristhorpe Bay.* Leckenby Collection.

A. woodwardi, H. G. Seeley, A. M. N. H. ser. 3, xvi (1865), pl. xi, f. 3, p. 236. *Cambridge Greensand; Cambridge.*

Ancyloceras hillsi (Sowerby), W. Keeping, Foss. Upware and Brickhill (1883), pl. ii, p. 91. *Lower Greensand; Upware.*

A. waltoni, J. Morris, A. M. N. H. xv (1845), pl. vi, f. 5, p. 33. *Inferior Oolite; Burton Bradstock.* Walton Collection.
Another specimen, G. F. Whidborne, Q. J. G. S. xxxix (1883), pl. xix, f. 1, p. 488. *Inferior Oolite; Burton Bradstock.* Presented by the Rev. G. F. Whidborne, M.A., F.G.S.

A. sp. 2, W. Keeping, Foss. Upware and Brickhill (1883), p. 153. *Lower Greensand* (derived); *Upware.*

Belemnites pistilliformis, Blainville, var. **subfusiformis**, d'Orbigny, W. Keeping, Foss. Upware and Brickhill (1883), pl. i, f. 6 *a*—*d*, p. 85. *Lower Greensand; Upware.*

B. subquadratus, Roemer, W. Keeping, Foss. Upware and Brickhill (1883), pl. i, f. 8 *a*, *b*, p. 86. *Lower Greensand; Brickhill.*

B. upwarensis, W. Keeping, Foss. Upware and Brickhill (1883), pl. i, f. 7 *a*—*d*, p. 87. *Lower Greensand; Upware.*

Beloteuthis leckenbyi, J. F. Blake, in R. Tate and J. F. Blake, The Yorkshire Lias (1876), pl. iv, f. 2 *a*, *b*, p. 314. *Lias (probably from the zone of A. serpentinus); Whitby.* Leckenby Collection.

Clymenia linearis, v. *C. undulata*.

Clymenia münsteri (Ansted), F. McCoy, Brit. Palæoz. Foss. (1852), pl. ii A, f. 12, 12 *a*, *b*, p. 402. *Endosiphonites münsteri*, D. T. Ansted, Trans. Camb. Phil. Soc. vi (1838), pl. viii, f. 1, p. 419. *Upper Devonian; South Petherwin.*

C. pattisoni, F. McCoy, A. M. N. H. ser. 2, viii (1851), p. 488, and Contrib. Brit. Pal. (1854), p. 239, and Brit. Palæoz. Foss. (1852), pl. ii A, f. 11, 11 *a*, *b*, p. 403. *Upper Devonian; South Petherwin.*

Clymenia quadrifera, F. McCoy, A. M. N. H. ser. 2, VIII (1851), p. 487, and Contrib. Brit. Pal. (1854), p. 238, and Brit. Palæoz. Foss. (1852), pl. ii A, f. 13, 13 *a*, *b*, p. 403. *Upper Devonian; South Petherwin.*

C. undulata (Münster). *C. linearis*, J. de C. Sowerby, Trans. Geol. Soc. ser. 2, v (1840), pl. liv, f. 19 *a*. *Endosiphonites carinatus*, D. T. Ansted, Trans. Camb. Phil. Soc. vi (1838), pl. viii, f. 2, p. 419. *E. minutus*, D. T. Ansted, *ibid.* (1838), pl. viii, f. 3, p. 420. *Upper Devonian ; South Petherwin.*

Conoceras llanvirnense, T. Roberts, Q. J. G. S. XL (1884) pl. xxviii, p. 636. *Llanvirn Beds ; N.W. of Llanvirn, Pembrokeshire.*

Crioceras occultus, H. G. Seeley, A. M. N. H. ser. 3, XVI (1865), pl. x, f. 1 *a*, *b*, *c*, p. 246. *(Drift?) Hunstanton.* Presented by Dr Cookson.

Cycloceras tenuiannulatum, v. Orthoceras tenuiannulatum.

Cyrtoceras (Meloceras) barrandei (Salter). *O. barrandei?*, J. F. Blake, Brit. Foss. Ceph. part i (1882), xviii, f. 12, p. 79. *Wenlock Shale ; Dudley.*

C. (Meloceras) corniculum, Barrande, J. F. Blake, Brit. Foss. Ceph. part i (1882), pl. xix, f. 8, 8 *a*, p. 173. *Wenlock Limestone ; Dudley.* Fletcher Collection.

C. intermedium, v. Cyrtoceras (Meloceras) m'coyi.

C. llandoveri, J. F. Blake, Brit. Foss. Ceph. part i (1882), pl. xxi, f. 1, 1 *a*, p. 171. *Upper Llandovery ; Craig-yr-Wyddon.*

C. (Meloceras) m'coyi, Foord. *C. intermedium* (McCoy), J. F. Blake, Brit. Foss. Ceph. part i (1882), pl. xxi, f. 2, 2 *a*, p. 179. *Phragmoceras intermedium*, F. McCoy, A. M. N. H. ser. 2, VII (1851), p. 45, and Contrib. Brit. Pal. (1854), p. 180, and Brit. Palæoz. Foss. (1852), p. 322. *Lower Ludlow ; Green Quarry, Leintwardine.*

C. multicameratum, Hall, F. McCoy, Brit. Palæoz. Foss. (1852), p. 312. J. F. Blake, Brit. Foss. Ceph. part i (1882), p. 184. *Lower Bala ; Knockdolian, Ayrshire.*

C. subornatum, F. McCoy, A. M. N. H. ser. 2, VIII (1851), p. 489, and Contrib. Brit. Pal. (1854), p. 240, and Brit. Palæoz. Foss. (1852), pl. ii A, f. 14, 14 a, p. 405. *Devonian ; Plymouth.*

Discites trochlea, F. McCoy, Brit. Palæoz. Foss. (1855), pl. iii H, f. 16, 16 a, p. 561. *Carboniferous Limestone ; Kendal.*

Endoceras proteiforme, v. Orthoceras araneosum ?

Endosiphonites carinatus, v. Clymenia undulata.

E. minutus, v. Clymenia undulata.

E. münsteri, v. Clymenia münsteri.

Gomphoceras æquale, Salter, Cat. (1873), p. 160.   J. F. Blake, Brit. Foss. Ceph. part i (1882), pl. xxvi, f. 6, 6 a, b, p. 188. *Wenlock Limestone; Dudley.* Fletcher Collection. The specimens in the Museum do not agree with the figures.

G. ellipticum (McCoy). *Poterioceras ellipticum,* F. McCoy, A. M. N. H. ser. 2, VII (1851), p. 45, and Contrib. Brit. Pal. (1854), p. 180, and Brit. Palæoz. Foss. (1852), p. 321. *Lower Ludlow ; Aymestry.* This specimen cannot be found.

G. neglectum, J. F. Blake, Brit. Foss. Ceph. part i (1882), pl. xxiii, f. 3, 3 a, b, p. 197. *Orthoceras (Poterioceras) pyriforme* (Sowerby), F. McCoy, Brit. Palæoz. Foss. (1852), p. 322. *Phragmoceras (Gomphoceras) pyriforme* (Sowerby), Salter, Cat. (1873), p. 174. *Lower Ludlow ; near Aymestry.*

G. ventricosum (Sowerby). *Phragmoceras ventricosum,* J. F. Blake, Brit. Foss. Ceph. part i (1882), pl. xxiv, f. 2, 2 a, p. 200. *May Hill Sandstone ; May Hill.*

Goniatites henslowi (Sowerby). *Ammonites henslowi,* J. Sowerby, Min. Conch. III (1821), pl. cclxii, p. 111. *Carboniferous Limestone ; Scarlet Point, Isle of Man.*

G. vinctus, J. de C. Sowerby, Trans. Geol. Soc. ser. 2, V (1840), pl. liv, f. 18. *Upper Devonian ; South Petherwin.*

Lituites anguiformis, v. Trocholites undosus.

Nautilus (Trocholites) anguiformis, v. Trocholites anguiformis.

Nautilus costato-coronatus, F. McCoy, A. M. N. H. ser. 2, XII (1853), p. 194, and Contrib. Brit. Pal. (1854), p. 263, and Brit. Palæoz. Foss. (1855), pl. iii H, f. 15*, p. 558. *Carboniferous Limestone ; Lowick.*

Nautilus tuberosus, F. McCoy, A. M. N. H. ser. 2, XII (1853), p. 195, and Contrib. Brit. Pal. (1854), p. 263, and Brit. Palæoz. Foss. (1855), pl. iii H, f. 15, 15 *a*, *b*, p. 562. *Carboniferous Limestone; Derbyshire.* Presented by W. Hopkins, Esq.

Onychoteuthis sagittata, v. Plesioteuthis sagittata.

Orthoceras cf. acuminatum, Eichwald, G. F. Whidborne, Devonian Fauna S. Eng. (1890), pl. xii, f. 7, p. 152. *Middle Devonian; Lummaton.*

O. angulatum, Wahlenberg. *O. bacchus*, Barrande, J. F. Blake, Brit. Foss. Ceph. part i (1882), pl. ix, f. 7, p. 111. *Lower Ludlow; Dudley.* Fletcher Collection.

O. angulatum, v. Orthoceras canaliculatum.

O. annulatum, Sowerby, var. **crassum**, Foord. *O. annulatum*, J. F. Blake, Brit. Foss. Ceph. part i (1882), pl. iv, f. 1, 5, p. 89. *Wenlock Limestone; Dudley.* Fletcher Collection.

O. araneosum?, Barrande, J. F. Blake, Brit. Foss. Ceph. part i (1882), p. 125. *Endoceras proteiforme*, Hall, H. A. Nicholson, G. M. IX (1872), p. 103, woodcut. *Skelgill Beds; Skelgill.* Presented by Prof. H. A. Nicholson.

O. arcuoliratum?, Hall, F. McCoy, Brit. Palæoz. Foss. (1852), p. 319. J. F. Blake, Brit. Foss. Ceph. part i (1882), pl. iii, f. 14, p. 84. *Middle Bala; Wrae Quarry, near Broughton.*

O. avelinei, Salter. *O. caereesiense*, H. Hicks, Q. J. G. S. XXXI (1875), pl. xi, f. 8, 9, p. 189. *Llanvirn Beds; Llanvirn Quarry.* Presented by Dr H. Hicks.

O. bacchus, v. Orthoceras angulatum.

O. baculiforme, J. W. Salter, Appendix to F. McCoy, Brit. Palæoz. Foss. (1855), p. vi. F. McCoy, *ibid.* (1852), pl. i L, f. 27, 27 *a*, p. 313. Salter, Cat. (1873), p. 187. J. F. Blake, Brit. Foss. Ceph. part i (1882), pl. xv, f. 3, 3 *a*, p. 160. *Upper Ludlow; Brigsteer, Westmoreland.*

O. barrandei, Salter, J. F. Blake, Brit. Foss. Ceph. part i (1882), pl. xviii, f. 10, p. 79. *Phragmoceras (Gomphoceras) liratum*, Salter, Cat. (1873), p. 174. *Lower Ludlow; Garden Quarry, Aymestry.*

Orthoceras barrandei?, v. Cyrtoceras (Meloceras) barrandei.

O. caereesiense, v. Orthoceras avelinei.

Orthoceras canaliculatum, Sowerby. *O. angulatum*, Wahlenberg, J. F. Blake, Brit. Foss. Ceph. part i (1882), pl. vii, f. 4, 9, p. 106. *Wenlock Limestone; Dudley.* Fletcher Collection.

O. cornu-vaccinum, F. McCoy, A. M. N. H. ser. 2, xii (1853), p. 196, and Contrib. Brit. Pal. (1854), p. 265, and Brit. Palæoz. Foss. (1855), pl. iii н, f. 17, 17 *a*, *b*, p. 568. *Carboniferous Limestone; Lowick.*

O. dimidiatum, Sowerby, J. F. Blake, Brit. Foss. Ceph. part i (1882), pl. vi, f. 11, p. 103. F. McCoy, Brit. Palæoz. Foss. (1852), p. 314. *O. dimidiatum*, var., Salter, Cat. (1873), p. 173. *Lower Ludlow; Leintwardine.*

O. elongatocinctum, Portlock, young specimen, J. F. Blake, Brit. Foss. Ceph. part i (1882), pl. xiii, f. 8, 8 *a*, p. 119. *Middle Bala; Barking.*

O. (Cycloceras) flemingi, F. McCoy, A. M. N. H. ser. 2, xii (1853), p. 196, and Contrib. Brit. Pal. (1854), p. 264, and Brit. Palæoz. Foss. (1855), pl. iii н, f. 18, 18 *a*, *b*, p. 569. *Carboniferous Limestone; Lowick.*

O. fluctuatum, Salter, Cat. (1873), p. 37. *O. reticinctum?*, J. F. Blake, Brit. Foss. Ceph. part i (1882), p. 122. *Llandeilo Beds; Wellfield, Builth.*

O. ibex, Sowerby, J. F. Blake, Brit. Foss. Ceph. part i (1882), pl. v, f. 5, p. 95. *Upper Ludlow; Benson Knot.*

O. ludense, Sowerby, J. F. Blake, Brit. Foss. Ceph. part i (1882), pl. x, f. 5, p. 156. *Lower Ludlow; Leintwardine.* The specimens do not agree well with the figure.

O. politum, F. McCoy, A. M. N. H. ser. 2, vii (1851), p. 46, and Contrib. Brit. Pal. (1854), p. 181, and Brit. Palæoz. Foss. (1852), pl. i L, f. 30, 30 *a*, p. 316. J. F. Blake, Brit. Foss. Ceph. part i (1882), pl. ix, f. 1, 2, 2 *a*, *b*, p. 141. *Bala Series; Glenwhapple.*

O. primævum (Forbes), J. F. Blake, Brit. Foss. Ceph. part i (1882), p. 148. *Creseis primæva*, E. Forbes, Q. J. G. S. i (1845), p. 146, woodcut 1. *Denbighshire Flags; Cefn-ddu.*

O. (Poterioceras) pyriforme, v. Gomphoceras neglectum.

Orthoceras sericeum, J. W. Salter, Appendix to A. C. Ramsay, Geol. North Wales (1866), pl. x, f. 4, 5, p. 356, and *ibid.* second edition (1881), p. 564. J. F. Blake, Brit. Foss. Ceph. part i (1882), pl. xiii, f. 2, p. 138. *Upper Tremadoc; Garth, Portmadoc.* Presented by D. Homfray, Esq.

     Orthoceras sp., H. Hicks, Q. J. G. S. xxix (1873), pl. iii, f. 27, p. 51. *Tremadoc Rocks; Tremaenhir, near Solva.* Presented by Dr H. Hicks.

O. striato-punctatum, Münster, var. **originale**, Barrande. *O. originale*, J. F. Blake, Brit. Foss. Ceph. part i (1882), pl. vii, f. 10, p. 110. *Wenlock Shale; Builth Bridge.*

O. striatulum, J. de C. Sowerby, Trans. Geol. Soc. ser. 2, v (1840), pl. liv, f. 20. *Upper Devonian; South Petherwin.*

O. subundulatum, Portlock. *Creseis sedgwicki*, E. Forbes, Q. J. G. S. i (1845), p. 146, woodcut 2. *Denbighshire Flags; Cefnddu, Denbighshire.*

O. tenuiannulatum (McCoy), J. F. Blake, Brit. Foss. Ceph. part i (1882), pl. v, f. 9, 9 *a*, p. 98. *Lower Ludlow; near Aymestry.*

     Cycloceras tenuiannulatum, F. McCoy, A. M. N. H. ser. 2, vii (1851), p. 46, and Contrib. Brit. Pal. (1854), p. 181, and Brit. Palæoz. Foss. (1852), pl. i L, f. 31, 31 *a*, p. 320. *Lower Ludlow; Green Quarry, Leintwardine.*

O. vagans, J. W. Salter, Appendix to F. McCoy, Brit. Palæoz. Foss. (1855), p. vi. F. McCoy, *ibid.* (1852), pl. i L, f. 28, 29, 29 *a*, p. 318. *Coniston Limestone; Coniston.*

O. sp., J. de C. Sowerby, Trans. Geol. Soc. ser. 2, v (1840), pl. liv, f. 21. *Upper Devonian; South Petherwin.*

O. sp., J. de C. Sowerby, Trans. Geol. Soc. ser. 2, v (1840), pl. liv, f. 22. *Upper Devonian; Beacon Cove, Mawgan.*

O. sp., Salter, Cat. (1873), p. 159. *Wenlock Limestone; Dudley.* Fletcher Collection.

O. sp., Salter, Cat. (1873), p. 187. *Upper Ludlow; High Thorns, Underbarrow.*

Phragmoceras intermedium, v. Cyrtoceras (Meloceras) m'coyi.

P. (Gomphoceras) liratum, v. Orthoceras barrandei.

Phragmoceras obliquum, J. F. Blake, Brit. Foss. Ceph. part i (1882), pl. xxiv, f. 7, 7 *a*, p. 203. *P. ventricosum*, Sowerby, *var.* 'dwarf with close septa,' Salter, Cat. (1873), p. 160. *Wenlock Limestone; Dudley.* Fletcher Collection.

P. ventricosum, v. Gomphoceras ventricosum.

P. ventricosum var., v. Phragmoceras obliquum.

Plesioteuthis sagittata (Münster). *Onychoteuthis sagittata*, G. G. Münster, Petrefactenkunde, heft vii (1846), pl. v, f. 5. *Solenhofen Slate; Eichstädt.* [Counterpart of the Type.] Münster Collection.

P. tricarinata (Münster). *Acanthoteuthis tricarinata*, G. G. Münster, Petrefactenkunde, heft vii (1846), pl. vi, f. 7, p. 60. *Solenhofen Slate; Solenhofen.* [Counterpart of the Type.] Münster Collection.

Poterioceras ellipticum, v. Gomphoceras ellipticum.

Scaphites meriani, Pictet and Campiche, A. J. Jukes-Browne, Q. J. G. S. xxxi (1875), pl. xiv, f. 1, 2, p. 287. *Cambridge Greensand; Cambridge.*

S. meriani, Pictet and Campiche, var. **simplex**, A. J. Jukes-Browne, Q. J. G. S. xxxi (1875), pl. xiv, f. 3, p. 287. *Cambridge Greensand; Cambridge.*

Teudopsis cuspidatus (Simpson), R. Tate and J. F. Blake, The Yorkshire Lias (1876), pl. iv, f. 3 *a, b,* p. 314. *Lias (probably zone of A. serpentinus); Whitby.* Leckenby Collection.

Toxoceras orbignyi, Baugier and Sauzé, G. F. Whidborne, Q. J. G. S. xxxix (1883), pl. xix, f. 2, p. 488. *Inferior Oolite; Half-way House, near Yeovil.* Presented by the Rev. G. F. Whidborne, M.A.

Trochoceras spurium, Salter, Cat. (1873), p. 160. *Wenlock Shale; Builth Bridge.*

Trocholites anguiformis, v. Trocholites undosus.

Trocholites undosus (Sowerby). *Lituites anguiformis*, J. W. Salter, Appendix to F. McCoy, Brit. Palæoz. Foss. (1855), p. viii. *Trocholites anguiformis*, F. McCoy, *ibid.* (1852), pl. i L, f. 26, p. 323. *Nautilus (Trocholites) anguiformis*, J. F. Blake, Brit.

Foss. Ceph. part i (1882), pl. xxviii, f. 2, 2 *a*, p. 213. *Middle Bala; Mynydd Fron Frys, near Chirk.*

Turrilites emericianus ?, d'Orbigny, A. J. Jukes-Browne, Q. J. G. S. XXXIII (1877), p. 493. *Cambridge Greensand; Cambridge.*

CRUSTACEA.

CIRRIPEDIA.

Turrilepas ketleyanus, Salter, Cat. (1873), p. 129. *Wenlock Limestone; Dudley.* Presented by C. Ketley, Esq.

Zoocapsa dilochorhamphia, H. G. Seeley, A. M. N. H. ser. 4, v (1870), p. 283. Portion of an *Avicula* or *Pecten*, H. Woodward, Cat. Brit. Foss. Crust. (1877), p. 146. *Lias; Lyme Regis.*

OSTRACODA.

Beyrichia complicata, Salter, F. McCoy, Brit. Palæoz. Foss. (1851), pl. i E, f. 3, 3 *a*, p. 136. *Llandeilo Beds; Pont-y-Meibion.*

B. devonica, T. R. Jones and H. Woodward, G. M. dec. 3, VI (1889), pl. xi, f. 3—5, p. 386. *Lower Devonian; New Cut, Torquay.*

B. klœdeni, F. McCoy, Brit. Palæoz. Foss. (1851), pl. i E, f. 2, p. 135. T. R. Jones, A. M. N. H. ser. 2, XVI (1855), p. 166. *Ludlow Beds; Cowan Head, Kendal.*

B. strangulata, v. Primitia strangulata.

Cythere aldensis (McCoy). *Cytheropsis aldensis*, F. McCoy, A. M. N. H. ser. 2, VIII (1851), p. 387, and Brit. Palæoz. Foss. (1852), pl. i L, f. 2. T. R. Jones and H. B. Holl, A. M. N. H. ser. 4, II (1868), p. 60. *Lower Bala; Aldens, Stinchar River.*

C.? umbonata, J. W. Salter, Appendix to F. McCoy, Brit. Palæoz. Foss. (1855), p. ii. *Ceratiocaris? umbonatus*, F. McCoy, Brit.

Palæoz. Foss. (1851), pl. i E, f. 6, p. 138. *Middle Bala; Dermydd-fawr.*

Cytheropsis aldensis, v. Cythere aldensis.

Echinocaris whidbornei, T. R. Jones and H. Woodward, G. M. dec. 3, VI (1889), pl. xi, f. 1, p. 385, and Rep. Brit. Assoc. (1889), p. 63. *Upper Devonian; Sloly, North Devon.*

Entomidella buprestis (Salter). T. R. Jones, Q. J. G. S. XXVIII (1872), pl. v, f. 15, p. 183. *Primitia buprestis,* Salter, Cat. (1873), p. 7. *Menevian Beds; Porth-y-rhaw, St Davids.* Presented by Dr H. Hicks.

E. marri, T. R. Jones and H. Woodward, Rep. Brit. Assoc. (1883), p. 215, and G. M. dec. 2, x (1883), p. 464. T. R. Jones, A. M. N. H. ser. 5, XIV (1884), pl. xv, f. 21, p. 401. *Upper Arenig; Nantlle Tramway, Pont Seiont.*

Entomis peregrina, G. F. Whidborne, Rep. Brit. Assoc. (1888), p. 681, and G. M. dec. 3, VI (1889), p. 29, and Devonian Fauna S. Eng. (1889), pl. iv, f. 15 *a, b, c,* p. 51. *Devonian; Lummaton.*

Leperditia vexata, H. Hicks, Q. J. G. S. XXVII (1871), p. 396. 'Larval Trilobite?', T. R. Jones, Q. J. G. S. XXVIII (1872), pl. v, f. 17, p. 184. *Menevian Beds; Porth-y-rhaw, St Davids.* Presented by Dr H. Hicks.

Primitia buprestis, v. Entomidella buprestis.

Primitia strangulata (Salter). *Beyrichia strangulata,* J. W. Salter, Appendix to F. McCoy, Brit. Palæoz. Foss. (1855), p. ii. F. McCoy, *ibid.* (1851), pl. i E, f. 1, 1 *a, b,* p. 136. *Coniston Limestone; Coniston Water Head.*

PHYLLOCARIDA.

Aptychopsis lapworthi, Woodward, T. R. Jones and H. Woodward, Brit. Assoc. Rep. (1884), p. 90. *Skelgill Beds; Skelgill.* Presented by J. E. Marr, Esq., F.R.S.

Peltocaris anatina, Salter, Cat. (1873), p. 93. T. R. Jones and H. Woodward, Rep. Brit. Assoc. (1884), p. 90. *Lower Wenlock; Rebecca Hill, Ulverston.*

Aptychopsis sp., T. R. Jones and H. Woodward, Rep. Brit.
Assoc. (1884), p. 92, and G. M. dec. 3, I (1884), p. 355.
Brathay Flags; Nanny Lane, Troutbeck. Presented by J. E.
Marr, Esq., M.A.

Caryocaris marri, H. Hicks, Q. J. G. S. XXXII (1876), p. 138.
T. R. Jones and H. Woodward, Rep. Brit. Assoc. (1883), p. 222.
Upper Arenig; Nantlle Tramway. Presented by J. E. Marr,
Esq., M.A.

Ceratiocaris ellipticus, v. Emmelezoe elliptica.

Ceratiocaris inornata [Salter MS.], F. McCoy, Brit. Palæoz. Foss.
(1851), pl. i E, f. 4, p. 137. Salter, Cat. (1873), p. 177. T. R.
Jones and H. Woodward, G. M. dec. 3, II (1885), p. 460, and
ibid. (1886), p. 459, and Rep. Brit. Assoc. (1885), p. 345, and
ibid. (1886), p. 232, and Brit. Palæoz. Phyllop. (1888), pl. x, f. 3,
p. 49. *Upper Ludlow; Benson Knot.*

C.? insperata, J. W. Salter, Appendix to A. C. Ramsay, Geol.
North Wales (1866), p. 295, woodcut 6, and *ibid.* second
edition (1881), p. 486. T. R. Jones and H. Woodward, Rep.
Brit. Assoc. (1885), p. 351, and G. M. dec. 3, II (1885), p. 466,
and Brit. Palæoz. Phyllop. (1888), p. 64. *Upper Tremadoc;
railway-cutting, Penmorfa.* Presented by D. Homfray, Esq.

C.? lata, J. W. Salter, Appendix to A. C. Ramsay, Geol. North
Wales (1866), p. 294, woodcut 4, and *ibid.* second edition (1881),
p. 486, woodcut 5. T. R. Jones and H. Woodward, Rep. Brit.
Assoc. (1885), p. 351, and G. M. dec. 3, II (1885), p. 465, and
Brit. Palæoz. Phyllop. (1888), p. 63. *Hymenocaris? latus,*
J. W. Salter, *ibid.* (1866), p. 240. *Upper Tremadoc; Garth,
Portmadoc.*

C. leptodactylus (McCoy), T. R. Jones and H. Woodward,
Rep. Brit. Assoc. (1885), p. 339, and G. M. dec. 3, II (1885),
p. 387, and Rep. Brit. Assoc. (1886), p. 229, and G. M. dec. 3,
III (1886), p. 456, and Brit. Palæoz. Phyllop. (1888), pl. vi,
f. 4 *a—c,* 5, p. 14. *Pterygotus leptodactylus,* F. McCoy, A. M.
N. H. ser. 2, IV (1849), p. 394, and Contrib. Brit. Pal. (1854),
p. 140, and Brit. Palæoz. Foss. (1851), pl. i E, f. 7, 7 *a, b* (not
f. 7 *c, d*), p. 175. *Leptocheles leptodactylus,* F. McCoy, Q. J.
G. S. IX (1853), p. 14. *Lower Ludlow; Leintwardine.*

Ceratiocaris? perornata [Salter MS.], T. R. Jones and H. Woodward, Rep. Brit. Assoc. (1885), p. 352, and G. M. dec. 3, II (1885), p. 466, and Brit. Palæoz. Phyllop. (1888), p. 64. *Upper Ludlow; Benson Knot, Kendal.*

C. robusta, Salter, T. R. Jones and H. Woodward, Rep. Brit. Assoc. (1885), p. 349, and G. M. dec. 3, II (1885), p. 464, and Rep. Brit. Assoc. (1886), p. 231, and G. M. dec. 3, III (1886), p. 457, and Brit. Palæoz. Phyllop. (1888), pl. xi, f. 8, 9, p. 44. *Pterygotus leptodactylus*, F. McCoy (*pars*), A. M. N. H. ser. 2, IV (1849), p. 394, and Contrib. Brit. Pal. (1854), p. 140, and Brit. Palæoz. Foss. (1851), pl. i E, f. 7 c, d, p. 175. *Lower Ludlow; Leintwardine.*

C. salteriana, T. R. Jones and H. Woodward, Rep. Brit. Assoc. (1885), p. 348, and G. M. dec. 3, II (1885), p. 462, and Brit. Palæoz. Phyllop. (1888), pl. vii, f. 2, p. 55. *Lower Ludlow; Dudley.* Fletcher Collection.

C. solenoides, F. McCoy, A. M. N. H. ser. 2, IV (1849), p. 413, and Contrib. Brit. Pal. (1854), p. 152, and Brit. Palæoz. Foss. (1851), pl. i E, f. 5, 5 a, p. 138. T. R. Jones and H. Woodward, Rep. Brit. Assoc. (1885), p. 347, and G. M. dec. 3, II (1885), p. 461, and Brit. Palæoz. Phyllop. (1888), pl. viii, f. 4 a, b, 5, 7 a, b, 10 a, b, 11 a—c, p. 52. *Upper Ludlow; Benson Knot, Kendal.*

C. stygia, Salter, **var.**, T. R. Jones and H. Woodward, Rep. Brit. Assoc. (1886), p. 232, and G. M. dec. 3, III (1886), p. 459. T. R. Jones and H. Woodward, Brit. Palæoz. Phyllop. (1888), pl. x, f. 7 a, b, p. 41. *C. stygia?, ibid.* pl. xi, f. 1, p. 41. *Upper Ludlow; Benson Knot, Kendal.*

Another specimen, T. R. Jones and H. Woodward, G. M. dec. 3, II (1885), p. 395, and Rep. Brit. Assoc. (1885), p. 344, and Brit. Palæoz. Phyllop. (1888), p. 40. *Denbighshire Series; Dinas-bran, Llangollen.*

C.? umbonatus, v. Cythere? umbonata.

C.? sp. (C. latus? or insperatus?), T. R. Jones and H. Woodward Brit. Assoc. Rep. (1883), p. 220, and *ibid.* (1888), woodcut 10 p. 178. *Lingula Flags; Wern, Portmadoc.*

C. sp., T. R. Jones and H. Woodward, G. M. dec. 3, V (1888),

p. 147, woodcut 8. *Upper Ludlow; near Beck Mills, Kendal.*

Discinocaris sp., T. R. Jones and H. Woodward, G. M. dec. 3, i (1884), p. 351. *Coniston Mudstone; Skelgill Beck, Ambleside.*

Dithyrocaris lateralis, F. McCoy, Brit. Palæoz. Foss. (1851), pl. iii i, f. 36, p. 181 (? tail spines of D. colei). *Carboniferous Limestone; Derbyshire.* Presented by W. Hopkins, Esq.

Emmelezoe elliptica (McCoy), T. R. Jones and H. Woodward, Rep. Brit. Assoc. (1886), p. 232, and G. M. dec. 3, iii (1886), p. 460, and Brit. Palæoz. Phyllop. (1888), pl. viii, f. 1 *a, b*, p. 68. *Ceratiocaris ellipticus*, F. McCoy, A. M. N. H. ser. 2, iv (1849), p. 413, and Contrib. Brit. Pal. (1854), p. 152, and Brit. Palæoz. Foss. (1851), pl. i E, f. 8, p. 137. J. W. Salter, A. M. N. H. ser. 3, v (1860), p. 157, and Cat. Camb. Sil. Foss. (1873), p. 178. T. R. Jones and H. Woodward, Rep. Brit. Assoc. (1885), p. 352, and G. M. dec. 3, ii (1885), p. 466. *Upper Ludlow; Benson Knot, Kendal.*

Hymenocaris vermicauda, Salter, T. R. Jones and H. Woodward Rep. Brit. Assoc. (1888), woodcuts 3, 4, 5, p. 178. *Middle Lingula Flags; Borth, Portmadoc.*

Lingulocaris lingulæcomes, J. W. Salter, Appendix to A. C. Ramsay, Geol. North Wales (1866), pl. x, f. 1, p. 294, and *ibid.* second edition (1881), p. 485. T. R. Jones and H. Woodward, Rep. Brit. Assoc. (1883), p. 223. *Upper Tremadoc; Garth.*

L. salteriana ?, T. R. Jones and H. Woodward, Rep. Brit. Assoc. (1888), woodcut 7, p. 176. *Lower Lingula Flags; Caen-y-Coed, near Maentwrog.*

Another specimen, T. R Jones and H. Woodward, Rep. Brit. Assoc. (1888), p. 177. *Brathay Flags; E. side of Long Sleddale.*

Peltocaris anatina ?, Salter, T. R. Jones and H. Woodward, Rep. Brit. Assoc. (1884), p. 93. *Skelgill Beds; W. side Long Sleddale.* Presented by J. E. Marr, Esq., F.R.S.

P. anatina, v. Aptychopsis lapworthi.

P. sp., T. R. Jones and H. Woodward, Rep. Brit. Assoc. (1884), p. 94. *Skelgill Beds; Skelgill.* Presented by J. E. Marr, Esq.

Saccocaris major (Salter), T. R. Jones and H. Woodward, Rep.

Brit. Assoc. (1888), woodcut 1, p. 176. *Hymenocaris major,* Salter, Cat. (1873), p. 7. T. R. Jones and H. Woodward, Rep. Brit. Assoc. (1883), pp. 220—1. *Lower Lingula Flags; Caen-y-Coed, near Maentwrog.* Presented by D. Homfray, Esq.

Saccocaris minor, T. R. Jones and H. Woodward, Rep. Brit. Assoc. (1890), p. 424, woodcuts 1—17. *Arenig Beds; Craig-yr-hyrddod, Arenig.* Presented by Professor Hughes.

TRILOBITA.

Acidaspis erinaceus, J. E. Marr and H. A. Nicholson, Q. J. G. S. XLIV (1888), pl. xvi, f. 11, 12, p. 723. *Skelgill Beds (zone of A. erinaceus); Torver Beck.* Presented by J. E. Marr, Esq.

A. hughesi, Salter, Cat. (1873), p. 93. *Coniston Grit; Casterton Low Fell, Kirkby Lonsdale.*

A. robertsi, G. F. Whidborne, Rep. Brit. Assoc. (1888), p. 681, and G. M. dec. 3, VI (1889), p. 29, and Devonian Fauna S. Eng. (1889), pl. i, f. 17, 17 *a,* p. 12. *Middle Devonian; Lummaton.*

Æglina boia, H. Hicks, in Salter, Cat. (1873), p. 23, and Q. J. G. S. XXXI (1875), pl. x, f. 9, 9 *a,* p. 185. *Middle Arenig; Whitesand Bay.* Presented by Dr H. Hicks.

Æ. hughesi, H. Hicks, Q. J. G. S. XXXII (1876), p. 138. *Arenig Beds; Nantlle Tramway.* Presented by J. E. Marr, Esq., M.A.

Æ. obtusicaudata, H. Hicks, Q. J. G. S. XXXI (1875), pl. x, f. 3, p. 185. *Llanvirn Beds; Llanvirn Quarry.* Presented by Dr H. Hicks.

Æ. sp., R. Etheridge, in J. C. Ward, Geol. N. Part Eng. Lake District (1876), pl. xii, f. 3, p. 111. J. Postlethwaite, Trans. Cumb. and Westm. Assoc. (1885), pl. ii, f. 12. J. Postlethwaite and J. G. Goodchild Proc. Geol. Assoc. IX (1886), pl. vii, f. 12, p. 461. *Skiddaw Slates; Great Knot, near Randel Crag.* Dover Collection.

Æ. sp., J. Postlethwaite, Trans. Cumb. and Westm. Assoc. (1885), pl. iii, f. 20, p. 74. J. Postlethwaite and J. G. Goodchild, Proc. Geol. Assoc. IX (1886), pl. viii, f. 20, p. 466. *Skiddaw Slates; S.W. of Randel Crag, Skiddaw.* Dover Collection.

Æglina sp., J. Postlethwaite, Trans. Cumb. and Westm. Assoc. (1885), pl. i, f. 6, p. 75. J. Postlethwaite and J. G. Goodchild, Proc. Geol. Assoc. IX (1886), pl. vi, f. 6, p. 458. *Skiddaw Slates; S.W. of Randel Crag, Skiddaw.* Dover Collection.

Æ. sp., J. Postlethwaite, Trans. Cumb. and Westm. Assoc. (1885), pl. iii, f. 18, p. 75. J. Postlethwaite and J. G. Goodchild, Proc. Geol. Assoc. IX (1886), pl. viii, f. 18, p. 465. *Skiddaw Slates; Buzzard Crag, Skiddaw.* Dover Collection.

Æ.? (or **Remopleurides**), J. Postlethwaite, Trans. Cumb. and Westm. Assoc. (1885), pl. ii, f. 11, p. 75. J. Postlethwaite and J. G. Goodchild, Proc. Geol. Assoc. IX (1886), pl. vii, f. 11, p. 461. *Skiddaw Slates; Buzzard Crag, Skiddaw.* Dover Collection.

Agnostus agnostiformis (McCoy). *Trinodus agnostiformis,* F. McCoy, Brit. Palæoz. Foss. (1851), pl. i E, f. 10, p. 141. *T. tardus,* F. McCoy (*non* Barrande), *ibid.* (1851), pl. i E, f. 9, p. 142. *Middle Bala; Rhiwlas.*

A. barrandei [Salter MS.], H. Hicks, Q. J. G. S. XXVIII (1872), pl. v, f. 5, p. 176. *Menevian Beds; Pen-pleidiau, St Davids.*

A. cambrensis, H. Hicks, Q. J. G. S. XXVII (1871), pl. xvi, f. 11, 11 a, 12, 12 a, p. 400. *Harlech Beds; Headland, near Nun's Well, St Davids.* Presented by Dr H. Hicks, F.R.S.
    Another specimen, H. Hicks, Q. J. G. S. XXVIII (1872), pl. v, f. 1. *Menevian Beds; Porth-y-rhaw, St Davids.*

A. davidis, Salter, H. Hicks, Q. J. G. S. XXVIII (1872), pl. v, f. 2, 4, p. 174. *Menevian Beds; Porth-y-rhaw, St Davids.*

A. dux, C. Callaway, Q. J. G. S. XXXIII (1877), pl. xxiv, f. 3, p. 665. *Shineton Shales (Tremadoc); Shineton.* [Counterpart of the Type.] Presented by Dr C. Callaway.

A. eskriggei, H. Hicks, Q. J. G. S. XXVIII (1872), pl. v, f. 7, p. 175. *Menevian Beds; Porth-y-rhaw, St Davids.*

A. hirundo, Salter, Cat. (1873), p. 22. H. Hicks, Q. J. G. S. XXXI (1875), pl. x, f. 10. *Middle Arenig; Whitesand Bay.*

A. m'coyi, Salter. *Diplorhina triplicata?,* Hawle and Corda, F. McCoy, Brit. Palæoz. Foss. (1851), pl. i E, f. 11, p. 142. *Llandeilo Beds; Pencerrig, near Builth.*

Agnostus scarabæoides, Salter, H. Hicks, Q. J. G. S. XXVIII (1872), pl. v, f. 8, p. 175. *Menevian Beds; Porth-y-rhaw, St Davids.*

A. scutalis, Salter, H. Hicks, Q. J. G. S. XXVIII (1872), pl. v, f. 9, 10, 12, p. 175. *Menevian Beds; Porth-y-rhaw, St Davids.*

Ampyx (Rhaphiophorus) aloniensis, J. E. Marr and H. A. Nicholson, Q. J. G. S. XLVI (1888), pl. xvi, f. 17, p. 724. *Skelgill Beds (zone of Ampyx aloniensis); Skelgill.* Presented by J. E. Marr, Esq., F.R.S.

A. latus, v. Ampyx nudus.

A. nudus (Murchison). *A. latus*, F. McCoy, A. M. N. H. ser. 2, IV (1849), p. 410, and Contrib. Brit. Pal. (1854), p. 148, and Brit. Palæoz. Foss. (1851), pl. i F, f. 13, 13 *a*, p. 147. *Llandeilo Beds; three miles N. of Builth.*

A. salteri, H. Hicks, in Salter, Cat. (1873), p. 22, and Q. J. G. S. XXXI (1875), pl. x, f. 7, p. 182. *Middle Arenig; slate quarry at Whitesand Bay.* Presented by Dr H. Hicks.

Anopolenus henrici, J. W. Salter, Q. J. G. S. XX (1864), pl. xiii, f. 4 *a, b*, p. 236, and *ibid.* XXI (1865), pp. 478, 481, woodcut 3. *Menevian Beds. Figs. 4 a, b, St Davids. Woodcut 3, Tafarnhelig, Maentwrog,* presented by D. Homfray, Esq.

A. impar, H. Hicks, Q. J. G. S. XXVIII (1872), pl. vii, f. 1—7, p. 179. *Menevian Beds; Porth-y-rhaw, St Davids*, presented by Dr H. Hicks. *Fig. 2 from Maentwrog?* presented by D. Homfray, Esq.

A. salteri, H. Hicks, Q. J. G. S. XXVIII (1872), pl. vii, f. 8, 9, 11, p. 179. *Menevian Beds; Porth-y-rhaw, St Davids.*

Arionellus longicephalus, H. Hicks, Q. J. G. S. XXVIII (1872), pl. v, f. 20, 22, p. 176. *Menevian Beds; Porth-y-rhaw, St Davids.* Presented by Dr H. Hicks.

Asaphus (Isotelus) affinis, F. McCoy, Brit. Palæoz. Foss. (1851), pl. i F, f. 3, p. 169. *Upper Tremadoc; over Iron-works, Tremadoc.*

A. (Isotelus) homfrayi, J. W. Salter, Appendix to A. C. Ramsay, Geol. North Wales (1866), pl. viii, f. 11, p. 311, and *ibid.* second edition (1881), p. 506, and Mon. Brit. Tril. (1866), pl. xxiv, f. 6, 8, 11, p. 165. *Tremadoc Beds; Tuhwnt-yr-bwlch, near Portmadoc.* Presented by D. Homfray, Esq.

Asaphus (Basilicus) laticostatus (McCoy, *pars*), J. W. Salter, Mon. Brit. Tril. (1866), pl. xviii, f. 6, p. 158. *Isotelus (Basilicus) laticostatus,* F. McCoy (*non* Green), Brit. Palæoz. Foss. (1851), pl. i E, f. 18 *a*, p. 170. *Llandeilo Flags; Maen Goran.*

A. menapiæ, v. Niobe menapiensis.

A. (Basilicus?) radiatus, J. W. Salter, Mon. Brit. Tril. (1866), pl. xviii, f. 2, p. 157. *Isotelus (Basilicus?) laticostatus* F. McCoy (*non* Green), Brit. Palæoz. Foss. (1851), pl. i E, f. 18, p. 170. *Middle Bala; Rhiwlas, North of Bala.*

A. solvensis, v. Niobe solvensis.

A. sp., R. Etheridge, in J. C. Ward, Geol. N. Part Eng. Lake District (1876), pl. xii, f. 1, p. 111. J. Postlethwaite, Trans. Cumb. and Westm. Assoc. (1885), pl. i, f. 1. Genus doubtful, J. Postlethwaite and J. G. Goodchild, Proc. Geol. Assoc. IX (1886), pl. vi, f. 1, p. 456. *Skiddaw Slates; N. side of Randel Crag, Skiddaw.* Dover Collection.

Barrandia cordai, F. McCoy, A. M. N. H. ser. 2, IV (1849), p. 409, and Contrib. Brit. Pal. (1854), p. 148, and Brit. Palæoz. Foss. (1851), pl. i F, f. 1, 1 *a*, p. 149. J. W. Salter, Mon. Brit. Tril. (1866), pl. xix, f. 5, 5 *a*, p. 142. *Llandeilo Beds; Pencerrig, near Builth.*

B. homfrayi, H. Hicks, Q. J. G. S. XXXI (1875), pl. ix, f. 8, p. 185. *Llanvirn Beds; Llanvirn Quarry.* Presented by Dr H. Hicks.

B. (Homalopteon) radians (McCoy). *Ogygia radians,* F. McCoy, A. M. N. H. ser. 2, IV (1849), p. 408, and Contrib. Brit. Pal. (1854), p. 147, and Brit. Palæoz. Foss. (1851), pl. i F, f. 2, 2 *a*, p. 149. *Llandeilo Beds; Pencerrig, near Builth.*

Brachymetopus ouralicus (De Verneuil), H. Woodward, Brit. Carb. Tril. (1884), pl. viii, f. 1, 3—6, p. 48. *Carboniferous Limestone; Settle.*

Calymene baylei, v. C. senaria.

Calymene blumenbachi, Brongniart. *C. subdiademata,* F. McCoy, Brit. Palæoz. Foss. (1851), pl. i F, f. 9, 9 *a*, 10, 10 *a*,

p. 166. *Fig.* 9, *Aymestry Limestone; Leintwardine. Fig.* 10, *Coniston Flags; Coniston Water Head.* These specimens cannot be found.

Calymene brevicapitata, v. C. cambrensis and C. senaria.

Calymene cambrensis, J. W. Salter, Mon. Brit. Tril. (1865), pl. ix, f. 14, p. 98. *Calymene brevicapitata,* Portlock, F. McCoy, Brit. Palæoz. Foss. (1851), pl. i F, f. 4, 5, p. 165. *Bala Beds; Nant-yr-Arian.*

C. hopkinsoni, H. Hicks, Q. J. G. S. xxxi (1875), pl. x, f. 4, 5, p. 187. *Llanvirn Beds; Llanvirn Quarry.* Presented by Dr H. Hicks.

C. parvifrons, J. W. Salter, Appendix to F. McCoy, Brit. Palæoz. Foss. (1855), p. iii. F. McCoy, *ibid.* (1851), pl. i F, f. 7, 7 *a,* p. 167. J. W. Salter, Mon. Brit. Tril. (1865), pl. ix, f. 25, and woodcut 22, p. 101. *Arenig Group; Tai-hirion.*

C. senaria, Conrad. *C. brevicapitata,* Portlock, F. McCoy, Brit. Palæoz. Foss. (1851), pl. i F, f. 6, p. 165. *Middle Bala; Applethwaite Common.*

    C. baylei, Barrande, F. McCoy, *ibid.* (1851), pl. i F, f. 8, 8 *a, b,* p. 165. *Llandeilo Beds; Tre-gib, Llandeilo.*

C. subdiademata, v. Calymene blumenbachi.

C. ultima, Salter, Cat. (1873), p. 22. *Arenig Beds; Ramsey Island.* Presented by Dr H. Hicks. This specimen cannot be found.

C. vexata, v. Neseuretus sp.

C. ?, F. McCoy, Brit. Palæoz. Foss. (1851), pl. i G, f. 11. *Middle Bala; Alt-yr-Anker.*

Carausia meneviensis, H. Hicks, Q. J. G. S. xxviii (1872), pl. vi, f. 7, p. 178. *Menevian Beds; Porth-y-rhaw, St Davids.*

Ceraurus clavifrons, v. Cheirurus juvenis and Sphærexochus boops.

C. octolobatus, v. Cheirurus octolobatus.

C. williamsi, v. Cheirurus bimucronatus.

Chasmops odini, v. Phacops conophthalmus.

Cheirurus bimucronatus (Murchison), J. W. Salter, Mon. Brit. Tril. (1864), pl. v, f. 4, p. 63. *Ceraurus williamsi,* F. McCoy, A. M. N. H. ser. 2, iv (1849), p. 408, and Contrib. Brit. Pal.

142   TRILOBITA.

(1854), p. 146, and Brit. Palæoz. Foss. (1851), pl. i F, f. 13, 13 *a, b*, p. 155. *Lower Llandovery; Goleugoed, Myddfai.*

Another specimen, J. W. Salter, Organic Remains, dec. VII (Mem. Geol. Survey, 1853), pl. ii, f. 5. *Wenlock Limestone; Dudley.* Fletcher Collection. The specimen in the collection does not agree well with the figure.

Cheirurus bimucronatus (Murchison) var. **acanthodes**, J. E. Marr and H. A. Nicholson, Q. J. G. S. XLIV (1888), pl. xvi, f. 7, 8, p. 722. *Skelgill Beds (zone of Phacops glaber); Skelgill.* Presented by J. E. Marr, Esq., F.R.S.

C. clavifrons, v. Cheirurus juvenis.

C. frederici, J. W. Salter, Appendix to A. C. Ramsay, Geol. North Wales (1866), p. 322, woodcut 10, and *ibid.* second edition (1881), p. 520. *Upper Tremadoc; Garth, Portmadoc.* Presented by D. Homfray, Esq.

C. juvenis, Salter. *C. clavifrons* (Dalman), J. W. Salter, Appendix to F. McCoy, Brit. Palæoz. Foss. (1855), p. iii. *Ceraurus clavifrons*, F. McCoy, *ibid.* (1851), pl. i G, f. 9, [9 *a* is missing], p. 154. *Middle Bala; Cefn-grugos, near Llanfyllin.*

C. (Pseudosphærexochus) moroides, J. E. Marr and H. A. Nicholson, Q. J. G. S. XLIV (1888), pl. xvi, f. 9, 10, 10 *a*, p. 722. *Skelgill Beds (zone of Phacops glaber); Skelgill.* Presented by J. E. Marr. Esq., F.R.S.

C. (Actinopeltis) octolobatus (McCoy). *Ceraurus octolobatus*, F. McCoy, A. M. N. H. ser. 2, IV (1849), p. 407, and Contrib. Brit. Pal. (1854), p. 146, and Brit. Palæoz. Foss. (1851), pl. i G, f. 10, 10 *a*, p. 154. *Middle Bala; Rhiwlas.*

C. pengellyi, G. F. Whidborne, G. M. dec. 3, VI (1889), p. 29, and Devonian Fauna S. Eng. (1889), pl. i, f. 11, 13, p. 8. *Middle Devonian; Lummaton.*

C. (Eccoptochile) sedgwicki (McCoy), J. W. Salter, Mon. Brit. Tril. (1864), pl. v, f. 17, p. 73. *Cryphæus sedgwicki*, F. McCoy, A. M. N. H. ser. 2, IV (1849), p. 406, and Contrib. Brit. Pal. (1854), p. 145. *Eccoptochile sedgwicki*, F. McCoy, Brit. Palæoz. Foss. (1851), pl. i F, f. 14, p. 155. *Llandeilo Beds; three miles North of Builth.*

Conocoryphe applanata, Salter, H. Hicks, Q. J. G. S. XXV (1869), pl. ii, f. 5, p. 53. *Menevian Beds; Porth-y-rhaw, St Davids*. Presented by Dr H. Hicks., F.R.S.

C. coronata (Barrande), H. Hicks, Q. J. G. S. XXVIII (1872), pl. vi, f. 11, p. 178. *Menevian Beds; Rhaiadr-du waterfall, Maentwrog*. Presented by D. Homfray, Esq.

C. homfrayi, Salter, Cat. (1873), p. 5. H. Hicks, Q. J. G. S. XXVIII (1872), pl. vi, f. 12, p. 178. *Menevian Beds; Waterfall Valley, near Maentwrog*. Presented by D. Homfray, Esq.

C. humerosa, J. W. Salter, Q. J. G. S. XXV (1869), pl. ii, f. 7, p. 54. *Menevian Beds; Porth-y-rhaw, St Davids*.

C. lyelli, H. Hicks, Q. J. G. S. XXVII (1871), pl. xvi, f. 2, 4, 7, p. 399. *Harlech Beds; Headland, near Nun's Well, St Davids*. Presented by Dr H. Hicks.

C. monile, Salter, Cat. (1873), p. 32. *Shineton Shales (Tremadoc); Shineton*.

C. solvensis, H. Hicks, Q. J. G. S. XXVII (1871), pl. xvi, f. 8, p. 400. *Harlech Beds; roadside between Whitchurch and Solva, St Davids*. Presented by Dr H. Hicks.

C.? verisimilis, J. W. Salter, Appendix to A. C. Ramsay, Geol. North Wales (1866), pl. vi, f. 13, p. 308, and *ibid.* second edition (1881), p. 503. *Lower Tremadoc; Penmorfa village*. Presented by D. Homfray, Esq.

C. williamsoni, Salter, Cat. (1873), p. 12. *Upper Lingula Flags; Rhiw-felyn, Upper Mawddach*. Presented by D. Homfray, Esq.

Cryphæus sedgwicki, v. Cheirurus sedgwicki.

Cybele punctata, v. Encrinurus punctatus.

Cybele rugosa (Portlock), J. W. Salter, Appendix to F. McCoy, Brit. Palæoz. Foss. (1855), p. iii. *Zethus rugosus*, F. McCoy, *ibid.* (1851), pl. i G, f. 8. *Coniston Limestone; Coniston*.

C. verrucosa (Dalman). *Zethus atractopyge*, F. McCoy, Brit. Palæoz. Foss. (1851), pl. i G, f. 1—5, p. 156. *Middle Bala; Ravenstone Dale, Alt-yr-Anker, and Coniston*.

Cyphaspis acanthina, C. Coignou, Q. J. G. S. XLVI (1890), p. 422, woodcut 5. *Carboniferous Limestone; Butterhaw, Cracoe*.

144 TRILOBITA.

Cyphaspis megalops (McCoy), J. W. Salter, Organic Remains, dec. VII (Mem. Geol. Survey, 1853), pl. v, f. 2, 7. *Wenlock Limestone; Dudley.* Fletcher Collection.
Another specimen, H. A. Nicholson and R. Etheridge jun., Mon. Sil. Foss. Girvan I, fascic. II (1880), p. 140. *May Hill Beds; Mulloch Quarry, near Girvan.*

Deiphon forbesi, Barrande, J. W. Salter, Mon. Brit. Tril. (1865), pl. vii, f. 11, p. 88. *Wenlock Limestone; Dudley.* Fletcher Collection.

Dikelocephalus (Centropleura) furca, J. W. Salter, Appendix to A. C. Ramsay, Geol. North Wales (1866), pl. vi, f. 4, p. 303, and *ibid.* second edition (1881), p. 497. *Lower Tremadoc; Penmorfa.* Presented by D. Homfray, Esq.

Diplorhina triplicata?, v. Agnostus m'coyi.

Dysplanus centrotus, v. Illænus bowmanni.

Eccoptochile sedgwicki, v. Cheirurus sedgwicki.

Encrinurus multiplicatus, Salter, Cat. (1873), p. 51. *Middle Bala; Barking, Dent.*

E. punctatus (Wahlenberg). *Cybele punctata*, T. W. Fletcher, Q. J. G. S. VI (1850), pl. xxxii, f. 2, 3, 5, p. 403. *Wenlock Shale; Malvern.* Fletcher Collection.

E. sexcostatus (Salter), J. W. Salter, Appendix to F. McCoy, Brit. Palæoz. Foss. (1855), p. iv. *Zethus sexcostatus*, F. McCoy, *ibid.* (1851), pl. i G, f. 6, 7, p. 157. *Middle Bala; Rhiwlas.*

Erinnys venulosa, Salter, H. Hicks, Q. J. G. S. XXVIII (1872), pl. vi, f. 1—4, 6, p. 177. *Menevian Beds; f.* 1—4 *from Porth-y-rhaw, St Davids,* presented by Dr H. Hicks. *Figs.* 5, 6, *from the Waterfall Valley, Maentwrog,* f. 6 presented by D. Homfray, Esq.

Griffithides acanthiceps, H. Woodward, Brit. Carb. Tril. (1883), pl. vi, f. 2, 10, pl. vii, f. 3, p. 32, and G. M. dec. 2, x (1883), p. 483. *Carboniferous Limestone; Settle.*

G. mesotuberculatus, v. Phillipsia eichwaldi *var.* mucronata.

G. moriceps, H. Woodward. Brit. Carb. Tril. (1884), pl. vii, f. 9—11, p. 39, and G. M. dec. 2, x (1883), p. 487. *Carboniferous Limestone; Settle.*

Griffithides seminiferus, H. Woodward, Brit. Carb. Tril. (1883), pl. v, f. 4, p. 28. *Carboniferous Limestone; Derbyshire.*

G. seminiferus ?, H. Woodward, Brit. Carb. Tril. (1884), pl. viii, f. 14, p. 28. *Carboniferous Limestone; Settle.* Burrow Collection.

Harpes angustus, J. E. Marr and H. A. Nicholson, Q. J. G. S. XLIV (1888), pl. xvi, f. 15, 16, 16 *a*, p. 724. *Skelgill Beds; Skelgill.* Fig. 15 from the zone of *Ampyx aloniensis*, fig. 16 from the zone of *Phacops glaber.* Presented by J. E. Marr, Esq., F.R.S.

H. judex, J. E. Marr and H. A. Nicholson, Q. J. G. S. XLIV (1888), pl. xvi, f. 13, 14, 14 *a*, p. 723. *Skelgill Beds; Skelgill.* Fig. 13 from the zone of *Ampyx aloniensis*, fig. 14 from the zone of *Phacops glaber.* Presented by J. E. Marr, Esq.

H. parvulus, F. McCoy, A. M. N. H. ser. 2, VIII (1851), p. 387, and Contrib. Brit. Pal. (1854), p. 209, and Brit. Palæoz. Foss. (1852), pl. i L, f. 3, p. 336 (footnote). *Middle Bala; Wrae Quarry, Upper Tweed.*

Holocephalina inflata, H. Hicks, Q. J. G. S. XXVIII (1872), pl. vi, f. 8—10, p. 178. *Menevian Beds; Porth-y-rhaw, St Davids.*

Homalonotus bisulcatus, J. W. Salter, Appendix to F. McCoy, Brit. Palæoz. Foss. (1855), p. v. F. McCoy, *ibid.* (1851), pl. i G, f. 24—31, p. 168 (f. 24, 25, 28 and 29 missing). J. W. Salter, Mon. Brit. Tril. (1865), pl. x, f. 7, 8 (missing), p. 105. *Middle Bala; Bryn Eithin, Acton Scott, and Ravenstone Dale.*

H. bisulcatus ?, J. W. Salter, Appendix to A. C. Ramsay, Geol. North Wales (1866), pl. xi A, f. 8, p. 328, and *ibid.* second edition (1881), p. 526. *Arenig Group; Ty-obry.* Presented by D. Homfray, Esq.

H. cylindricus, J. W. Salter, Mon. Brit. Tril. (1865), woodcuts 27, 28, pp. 116—7. *Woolhope Limestone; Woolhope.* Fig. 28 (from the Strickland Collection) is missing.

H. goniopygæus, H. Woodward, G. M. dec. 2, IX (1882), pl. iv, f. 1, p. 157. *Middle Devonian; Smuggler's Cove, Torquay.* Presented by E. B. Tawney, Esq., M.A.

H. monstrator, v. Neseuretus sp.

Homalonotus rudis, J. W. Salter, Appendix to F. McCoy, Brit.
Palæoz. Foss. (1855), p. v.   F. McCoy, *ibid.* (1851), pl. i E, f. 2
0, 20 *a*, p. 168.   *Middle Bala; Capel Garmon, Denbighshire.*

H. sedgwicki, J. W. Salter, Mon. Brit. Tril. (1865), p. 107,
woodcut 25.   *Middle Bala; Ravenstone Dale, and Llanwddyn.*

Illænopsis? acuticaudata, H. Hicks, Q. J. G. S. xxxi (1875),
pl. ix, f. 5, p. 184.   *Llanvirn Beds; Llanvirn Quarry.* Presented by Dr H. Hicks.

Illænus (Dysplanus) bowmanni, J. W. Salter, Mon. Brit. Tril.
(1867), pl. xxviii, f. 10, p. 185.   *Dysplanus centrotus*, F. McCoy,
Brit. Palæoz. Foss. (1851), pl. i E, f. 19, 19 *a*, p. 173 (from
Llanwddyn).   *Illœnus latus*, F. McCoy, A. M. N. H. ser. 2, iv
(1849), p. 404, and Contrib. Brit. Pal. (1854), p. 143, and Brit.
Palæoz. Foss. (1851), pl. i E, f. 17, 17 *a*, p. 172.   J. W. Salter,
Mon. Brit. Tril. (1867), pp. 186, 215, woodcut 55.   H. A.
Nicholson and R. Etheridge, jun., Mon. Sil. Foss. Girvan, i,
fascic. ii (1880), p. 156.   *Middle Bala; Llanwddyn, and Wrae
Quarry, near Broughton.*

I. hughesi, H. Hicks, Q. J. G. S. xxxi (1875), pl. ix, f. 7, p. 184.
Llanvirn Beds; Llanvirn Quarry.   Presented by Dr H. Hicks.

I. latus, v. Illænus bowmanni.

I. (Panderia) lewisi, J. W. Salter, Mon. Brit. Tril. (1867), pl. xxvi,
f. 2, 2 *a*, *b*, p. 183.   *Bala Limestone; Moelydd, Oswestry.*
Presented by the Rev. D. P. Lewis, M.A.

I. rosenbergi, Eichwald, F. McCoy, Brit. Palæoz. Foss. (1851),
pl. i G, f. 33—35, p. 172.   J. W. Salter, Mon. Brit. Tril. (1867),
pl. xxix, f. 2, 3, 4, p. 199.   *Coniston Limestone; Sunny Brow,
Coniston.*

I. (Dysplanus) thomsoni, J. W. Salter, Mon. Brit. Tril. (1867),
pl. xxx, f. 10, p. 188.   *Llandovery Beds; Mulloch, Girvan.*

I. (Bumastus) sp., J. W. Salter, Mon. Brit. Tril. (1883), p. 215,
woodcut 57, and Cat. (1873), p. 54.   *Middle Bala; Mynydd
Fron Frys, Chirk.*

Isotelus laticostatus, v. Asaphus radiatus and A. laticostatus.

Lichas anglicus (Beyrich).   *L. bucklandi*, T. W. Fletcher, Q.
J. G. S. vi (1850), pl. xxvii, f. 2, 3, 3 *a*, xxvii *bis*, f. 1, 1 *a*,
p. 235.   *Wenlock Limestone; Dudley.*   Fletcher Collection.

Lichas barrandei, T. W. Fletcher, Q. J. G. S. VI (1850), pl. xxvii, f. 10, xxvii *bis*, f. 5, p. 238. *Wenlock Limestone; Dudley.* Fletcher Collection.

L. bucklandi, v. Lichas anglicus.

L. grayi, T. W. Fletcher, Q. J. G. S. VI (1850), pl. xxvii, f. 8, xxvii *bis*, f. 3, 3 *a*, *b*, p. 287. *Wenlock Limestone; Dudley.* Fletcher Collection.

L. hirsutus, T. W. Fletcher, Q. J. G. S. VI (1850), pl. xxvii, f. 6, 6 *a*, 7, 7 *a*, xxvii *bis*, f. 2, 2 *a*, p. 236. *Wenlock Limestone; Dudley.* Fletcher Collection.

L. laciniatus, Dalman. *L. subpropinqua*, F. McCoy, Brit. Palæoz. Foss. (1851), pl. i F, f. 17, 17 *a*, p. 150. *Coniston Limestone; Coniston.*

L. nodulosus, J. W. Salter, Appendix to F. McCoy, Brit. Palæoz. Foss. (1855), p. iv. *Trochurus nodulosus*, F. McCoy, *ibid.* (1851), pl. i F, f. 16, p. 151. *Middle Bala; Pont-y-Glyn-Diffwys, Corwen.*

L. salteri, T. W. Fletcher, Q. J. G. S. VI (1850), pl. xxvii, f. 9, 9 *a*, p. 237. *Wenlock Limestone; Dudley.* Fletcher Collection.

L. scutalis, v. Lichas verrucosus.

L. subpropinqua, v. Lichas laciniatus.

L. verrucosus (Eichwald). *L. scutalis*, Salter, Cat. (1873), p. 130. *Wenlock Shale; Malvern.* Fletcher Collection.

Microdiscus punctatus, H. Hicks, Q. J. G. S. XX (1864), pl. xiii, f. 11 *c*, p. 237. *Menevian Beds; Porth-y-rhaw, St Davids.* Presented by Dr H. Hicks.

M. sculptus, H. Hicks, Q. J. G. S. XXVII (1871), pl. xvi, f. 9, 10, p. 400. *Harlech Beds; Headland, near Nun's Well, St Davids.* Presented by Dr H. Hicks.

Neseuretus? elongatus, H. Hicks, Q. J. G. S. XXIX (1873), pl. iii, f. 2, 3, p. 45. *Tremadoc Beds; Tremaenhir.* Presented by Dr H. Hicks.

N.? elongatus, Hicks, var. **obesus**, H. Hicks, Q. J. G. S. XXIX (1873), pl. iii, f. 4. *Tremadoc Beds; Ramsey Island.* Presented by Dr H. Hicks.

Neseuretus quadratus, H. Hicks, Q. J. G. S. xxix (1873), pl. iii, f. 11, 13, 23, 25, p. 45. *Tremadoc Beds; Ramsey Island.* Presented by Dr H. Hicks.

N. ramseyensis, H. Hicks, Q. J. G. S. xxix (1873), pl. iii, f. 8, 9, 15, 16, 18, 19, 22, p. 44. *Tremadoc Beds; Ramsey Island and Tremaenhir.* Presented by Dr H. Hicks.

N. recurvatus, H. Hicks, Q. J. G. S. xxix (1873), pl. iii, f. 5 a, 6, p. 45. *Tremadoc Beds; Ramsey Island and Tremaenhir.* Presented by Dr H. Hicks.

N. sp. *Homalonotus monstrator*, Salter, Cat. (1873), p. 22. *Tremadoc Beds; Ramsey Island, St Davids.*

N. sp. *Calymene vexata*, Salter, Cat. (1873), p. 22. *Tremadoc Beds; Ramsey Island.*

Niobe doveri, R. Etheridge, in J. C. Ward, Geol. N. Part Eng. Lake District (1876), pl. xii, f. 2, p. 110. J. Postlethwaite, Trans. Cumb. and Westm. Assoc. (1885), pl. ii, f. 13. *Niobe?*, J. Postlethwaite and J. G. Goodchild, Proc. Geol. Assoc. ix (1886), pl. vii, f. 13, p. 461. *Skiddaw Slates; Brunstock Scar, below Randel Crag.* Dover Collection.

N. homfrayi, J. W. Salter, Mon. Brit. Tril. (1865), pl. xx, f. 3, 4, 9, p. 143, and Appendix to A. C. Ramsay, Geol. North Wales (1866), pl. vi, f. 5, p. 314, and *ibid.* second edition (1881), p. 511. *Lower Tremadoc; Penmorfa.* Presented by D. Homfray, Esq.

N. menapiensis, H. Hicks, Q. J. G. S. xxix (1873), pl. iv, f. 1, 7, 8, p. 46. *Asaphus menapiæ*, H. Hicks, in Salter, Cat. (1873), p. 23. *Tremadoc Beds; Ramsey Island.* Presented by Dr H. Hicks.

N. solvensis, H. Hicks, Q. J. G. S. xix (1873), pl. iv, f. 11—16, p. 47. *Asaphus solvensis*, H. Hicks, in Salter, Cat. (1873), p. 23. *Tremadoc Beds; Ramsey Island and Tremaenhir.* Presented by Dr H. Hicks.

Odontochile obtusi-caudata, v. Phacops obtusi-caudatus.

O. truncato-caudata, v. Phacops macrura.

Ogygia bullina, J. W. Salter, Mon. Brit. Tril. (1867), pl. xxv*, f. 5, p. 178. *Middle Arenig; Whitesand Bay.*

Ogygia peltata, J. W. Salter, Mon. Brit. Tril. (1867), pl. xxv*, f. 1, 2, 4, p. 177. *Middle Arenig; Whitesand Bay.*

O. radians, v. Barrandia radians.

Olenus (Sphærophthalmus) expansus, Salter, Cat. (1873), p. 12. *Upper Lingula Flags; Moel Gron.*

O. planti, Salter, Cat. (1873), p. 11. *Upper Lingula Flags; Moel Gron.* Presented by J. Plant, Esq.

Paradoxides harknessi, H. Hicks, Q. J. G. S. xxvii (1871), pl. xv, f. 9, 10, 11, p. 399. *Harlech Beds; St Davids.* Presented by Dr H. Hicks.

P. hicksi, Salter, H. Hicks, Q. J. G. S. xxv (1869), pl. iii, f. 1 *b*, 2, 10 *a*, p. 55. *Menevian Beds; Pen-pleidiau, St Davids.* Presented by Dr H. Hicks.

Phacops (Acaste) alifrons, J. W. Salter, Appendix to F. McCoy, Brit. Palæoz. Foss. (1855), p. ii. F. McCoy, *ibid.* (1851), pl. i G, f. 12—14, p. 159. J. W. Salter, Mon. Brit. Tril. (1864), pl. i, f. 31, 32, p. 33. *Middle Bala; Capel Garmon, and Pont-y-Glyn.*

P. (Acaste) apiculatus, J. W. Salter, Appendix to F. McCoy, Brit. Palæoz. Foss. (1855), p. iii. *Portlockia?* apiculata, F. McCoy, *ibid.* (1851), pl. i G, f. 17--19, p. 162. *Middle Bala; S. W. of Pwllheli, Glyn Ceiriog, and Rhiwargor.*

P. (Chasmops) conophthalmus (Boeck ?), J. W. Salter, Organic Remains, dec. vii, art. 1, (Mem. Geol. Survey, 1853), p. 11, and Mon. Brit. Tril. (1864), pl. iv, f. 25, pl. vi, f. 25, p. 40. *Chasmops odini?* (Eichwald), F. McCoy, Brit. Palæoz. Foss. (1851), pl. i G, f. 22, 23, p. 164. *Middle Bala; Llansantffraid.*

P. (Acaste) downingiæ (Murchison), J. W. Salter, Organic Remains, dec. vii (Mem. Geol. Survey, 1853), pl. i, f. 2, 3. *Wenlock Limestone; Dudley.* Fletcher Collection. These specimens cannot be identified.

P. (Acaste) downingiæ (Murchison) var. **cuneatus**, J. W. Salter, Mon. Brit. Tril. (1864), p. 28, woodcut 8. *Denbighshire Grits; Moel Seisiog, Llanrwst.*

P. elegans (Boeck and Sars), J. E. Marr and H. A. Nicholson, Q. J. G. S. xliv (1888), pl. xvi, f. 1—3, p. 720. *Skelgill Beds; figs. 1, and 3, zone of Phacops glaber, Skelgill : fig. 2, zone* *of Acidaspis erinaceus, Torver Beck.* Presented by J. E. Marr, Esq.

Phacops elegans (Boeck and Sars) var. **glaber**, J. E. Marr and H. A. Nicholson, Q. J. G. S. XLIV (1888), pl. xvi, f. 4, p. 721. *Skelgill Beds (glaber-zone); Skelgill.* Presented by J. E. Marr, Esq.

P. granulatus (Münster). *Calymene sp.*, J. de C. Sowerby, Trans. Geol. Soc. ser. 2, v (1840), pl. liv, f. 23, 24. *Portlockia granulata*, F. McCoy, Brit. Palæoz. Foss. (1851), p. 177. G. F. Whidborne, Dev. Fauna S. Eng. (1889), p. 4. *Devonian; Petherwin.*

P. llanvirnensis, H. Hicks, Q. J. G. S. XXXI (1875), pl. ix, f. 4, p. 187. *Llanvirn Beds; Llanvirn Quarry.* Presented by Dr H. Hicks.

P. (Chasmops) macrura, Angelin. *Odontochile truncato-caudatus* (Portlock), F. McCoy, Brit. Palæoz. Foss. (1851), pl. i G, f. 20, 21, p. 162. *Middle Bala; the Hollies, Church Stretton: and Coniston Limestone; Coniston.*

P. (Dalmanites) mucronatus (Brongniart), J. E. Marr and H. A. Nicholson, Q. J. G. S. XLIV (1888), pl. xvi, f. 5, 6, p. 721. *Skelgill Beds (zone of Ampyx aloniensis); Browgill.* Presented by J. E. Marr, Esq.

P. (Odontochile) obtusi-caudatus, J. W. Salter, Appendix to F. McCoy, Brit. Palæoz. Foss. (1855), p. ii. F. McCoy, *ibid.* (1851), pl. i G, f. 15, 16, p. 161. *Calcareous Flags; Coldwell.*

Another specimen, J. W. Salter, Mon. Brit. Tril. (1864), pl. i, f. 42 *a, b*, p. 45. *Coniston Flags; Coniston.*

These specimens cannot be found.

Phillipsia derbiensis (Martin), H. Woodward, Brit. Carb. Tril. (1883), pl. i, f. 3, p. 12. *Carboniferous Limestone; Settle.* Burrow Collection.

P. eichwaldi (Fischer) var. **mucronata**, McCoy. *Griffithides meso-tuberculatus*, F. McCoy, A. M. N. H. ser. 2, IV (1849), p. 406, and Contrib. Brit. Pal. (1854), p. 144, and Brit. Palæoz. Foss. (1851), pl. iii D, f. 10, 11, 11 *a*, p. 182. H. Woodward, Brit. Carb. Tril. (1883), p. 26. *Carboniferous Limestone; Derbyshire.* Presented by W. Hopkins, Esq.

Phillipsia gemmulifera (Phillips), H. Woodward, Brit. Carb. Tril. (1883), pl. iii, f. 4, p. 17. *Carboniferous Limestone; Derbyshire.*

P. truncatula (Phillips), H. Woodward, Brit. Carb. Tril. (1883), pl. iii, f. 10—12, p. 21. *Carboniferous Limestone; Hook Head, co. Wexford.*

Placoparia cambrensis, H. Hicks, Q. J. G. S. xxxi (1875), pl. ix, f. 1, p. 186. *Llanvirn Beds; Llanvirn Quarry.* Presented by Dr H. Hicks.

Plutonia sedgwicki, H. Hicks, Q. J. G. S. xxvii (1871), pl. xv, f. 1, 3, 4, 5, 7, p. 399. *Harlech Beds; Headland, near Nun's Well, St Davids.* Presented by Dr H. Hicks.

Portlockia? apiculata, v. Phacops apiculatus.

P. granulata, v. Phacops granulatus.

Proëtus brachypygus, J. E. Marr, and H. A. Nicholson, Q. J. G. S. xliv (1888), pl. xvi, f. 18, 19, p. 725. *Skelgill Beds (zone of Ampyx aloniensis); f. 18 from Skelgill, f. 19 from Pull Beck.* Presented by J. E. Marr, Esq., F.R.S.

P. champernowni, G. F. Whidborne, Devonian Fauna S. Eng. (1889), pl. ii, f. 13, p. 23. *Middle Devonian; Lummaton, near Torquay.*

P. fletcheri, Salter, Cat. (1873), p. 134. *Wenlock Limestone; Dudley.* Fletcher Collection.

Psilocephalus inflatus, J. W. Salter, Appendix to A. C. Ramsay Geol. North Wales (1866), p. 316, woodcut 8 a, and *ibid.* second edition (1881), p. 512, and Mon. Brit. Tril. (1866), p. 176, woodcut 41 a. *Lower Tremadoc; Penmorfa Road.* Presented by D. Homfray, Esq.

P. innotatus, J. W. Salter, Appendix to A. C. Ramsay, Geol. North Wales (1866), pl. vi, f. 9, 9 a, 10, 12, p. 315, and *ibid.* second edition (1881), p. 511, and Mon. Brit. Tril. (1866), pl. xx, f. 14, 17, 19, p. 175. *Lower Tremadoc; near Borth, Portmadoc.* Presented by D. Homfray, Esq.

Sphærexochus boops, J. W. Salter, Mon. Brit. Tril. (1864), pl. vi, f. 28, p. 79. *Ceraurus clavifrons* (Dalman), F. McCoy, Brit.

Palæoz. Foss. (1851), pl. i F, f. 12, p. 154. *Coniston Limestone; Applethwaite Common.*

Sphærexochus mirus, Beyrich, J. W. Salter, Mon. Brit. Tril. (1864), pl. vi, f. 6 *g*, p. 76. *Wenlock Limestone; Malvern.* Fletcher Collection.

Staurocephalus murchisoni, Barrande, F. McCoy, Brit. Palæoz. Foss. (1851), pl. i F, f. 15, 15 *a*, p. 153. J. W. Salter, Mon. Brit. Tril. (1865), pl. vii, f. 19, p. 84. *Middle Bala; Rhiwlas.*

Tretaspis fimbriatus, v. Trinucleus fimbriatus.

Trinodus agnostiformis, v. Agnostus agnostiformis.

T. tardus, v. Agnostus agnostiformis.

Trinucleus concentricus (Eaton). *T. gibbifrons*, F. McCoy, A. M. N. H. ser. 2, IV (1849), p. 411, and Contrib. Brit. Pal. (1854), p. 149, and Brit. Palæoz. Foss. (1851), pl. i E, f. 14, 14 *a*, p. 145. *Llandeilo Beds; Golden Grove, Llandeilo.* This specimen cannot be found.

T. etheridgei, H. Hicks, Q. J. G. S. XXXI (1875), pl. ix, f. 6, p. 182. *Llanvirn Beds; Llanvirn Quarry.* Presented by Dr H. Hicks.

T. fimbriatus, Murchison. *Tretaspis fimbriatus*, F. McCoy, Brit. Palæoz. Foss. (1851), pl. i E, f. 16, 16 *a*, p. 146. *Llandeilo Beds; Pencerrig, near Builth.*

T. gibbifrons, v. Trinucleus concentricus.

T. ramsayi, H. Hicks, Q. J. G. S. XXXI (1875), pl. x, f. 2, p. 183. *Upper Arenig; Porth-hayog.* Presented by Dr H. Hicks.

Trochurus nodulosus, v. Lichas nodulosus.

Zethus atractopyge, v. Cybele verrucosa.

Z. rugosus, v. Cybele rugosa.

Z. sexcostatus, v. Encrinurus sexcostatus.

EURYPTERIDA.

Eurypterus cephalaspis (Salter), J. W. Salter, Appendix to
F. McCoy, Brit. Palæoz. Foss. (1855), p. v. F. McCoy, *ibid.*
(1851), pl. i E, f. 21, p. 175. *Upper Ludlow; Kirkby Moor,
Kendal.*

Bellinurus? sp., F. McCoy, Brit. Palæoz. Foss. (1851), pl. i E,
f. 12. H. Woodward, Mon. Merostomata (1878), p. 238. *Hemiaspis aculeatus*, Salter, Cat. (1873), p. 178. *Upper Ludlow;
High Thorns, Underbarrow.*

Hemiaspis aculeatus, v. Bellinurus? sp.

Leptocheles leptodactylus, v. Ceratiocaris leptodactylus.

Pterygotus leptodactylus, v. Ceratiocaris robusta, and C. leptodactylus.

Pterygotus problematicus, Agassiz, J. W. Salter, Q. J. G. S.
VIII (1852), pl. xxi, f. 1 *a, b,* 2 *a, b,* p. 386. *Upper Ludlow;
Hagley Park; Herefordshire.* Strickland Collection.

ISOPODA.

Palæga carteri, H. Woodward, G. M. VII (1870), pl. xxii, f. 4,
p. 496. *Cambridge Greensand; Cambridge.*

P. m'coyi, J. Carter, G. M. dec. III, VI (1889), pl. vi, f. 1—7,
p. 193. *Cambridge Greensand; Cambridge.*

DECAPODA.

Archæocarabus bowerbanki, F. McCoy, A. M. N. H. ser. 2, IV
(1849), p. 174, and Contrib. Brit. Pal. (1854), p. 129. *London
Clay; Sheppey.*

Dromilites lamarcki (Desmarest). *Basinotopus lamarcki*, F.
McCoy, A. M. N. H. ser. 2, IV (1849), p. 168, and Contrib.
Brit. Pal. (1854), p. 123. *London Clay; Sheppey.*

Enoploclytia brevimana, F. McCoy, A. M. N. H. ser. 2, IV (1849), p. 332, and Contrib. Brit. Pal. (1854), p. 137. *Lower Chalk; Cherry Hinton.*

Eryma babeaui, Etallon, J. Carter, Q. J. G. S. XLII (1886), p. 548. *Oxford Clay; St Ives.* Presented by T. J. George, Esq., F.G.S.

E. georgei, J. Carter, Q. J. G. S. XLII (1886), pl. xvi, f. 4, 4 *a*, p. 549. *Oxford Clay; St Ives.* Presented by T. J. George, Esq., F.G.S.

E. mandelslohi (Meyer), J. Carter, Q. J. G. S. XLII (1886), pl. xvi, f. 2, 2 *a, b, c*, p. 546. *Oxford Clay; St Ives.* Presented by T. J. George, Esq.

E.? pulchella, J. Carter, Q. J. G. S. XLII (1886), pl. xvi, f. 5, 5 *a*, p. 550. *Oxford Clay; St Ives.* Presented by T. J. George, Esq.

E. ventrosa (Meyer), J. Carter, Q. J. G. S. XLII (1886), p. 547. *Oxford Clay; St Ives.* Presented by T. J. George, Esq.

E. villersi, Morière, J. Carter, Q. J. G. S. XLII (1886), pl. xvi, f. 3, 3 *a*, p. 548. *Oxford Clay; St Ives.* Presented by T. J. George, Esq.

Eryon barrovensis, F. McCoy, A. M. N. H. ser. 2, IV (1849), p. 172, and Contrib. Brit. Pal. (1854), p. 127. *Lias; Barrow-on-Soar.*

E. sublevis, J. Carter, Q. J. G. S. XLII (1886), pl. xvi, f. 1, p. 545. *Oxford Clay; St Ives.*

Glyphæa cretacea, F. McCoy, A. M. N. H. ser. 2, XIV (1854), pl. iv, f. 2, p. 118, and Contrib. Brit. Pal. (1854), p. 268, plate, f. 2, 2 *a*. *Cambridge Greensand; Cambridge.*

G. hispida, J. Carter, Q. J. G. S. XLII (1886), pl. xvi, f. 6, 6 *a*, p. 550. *Oxford Clay; St Ives.* Presented by T. J. George, Esq.

G. regleyana, Meyer, J. Carter, Q. J. G. S. XLII (1886), p. 551. *Oxford Clay; St Ives.* Presented by T. J. George, Esq.

Goniochirus cristatus, J. Carter, Q. J. G. S. XLII (1886), pl. xvi, f. 9, 9 *a*, p. 555. *Oxford Clay; St Ives.* Presented by T. J. George, Esq.

G. platycheles (McCoy). *Pagurus? platycheles*, F. McCoy, A.

M. N. H. ser. 2, IV (1849), p. 171, and Contrib. Brit. Pal. p. 126. *Great Oolite; Minchinhampton.*

Hoploparia belli, F. McCoy, A. M. N. H. ser. 2, IV (1849), p. 178, and Contrib. Brit. Pal. (1854), p. 133. *London Clay; Sheppey.*

H. prismatica, F. McCoy, A. M. N. H. ser. 2, IV (1849), p. 176, and Contrib. Brit. Pal. (1854), p. 131. *Speeton Clay; Speeton.*

Magila dissimilis, J. Carter, Q. J. G. S. XLII (1886), pl. xvi, f. 8, 8 *a*, p. 552. *Oxford Clay; St Ives.* Presented by T. J. George, Esq., F.G.S.

M. levimana, J. Carter, Q. J. G. S. XLII (1886), pl. xvi, f. 7, 7 *a*, p. 552. *Oxford Clay; St Ives.* Presented by T. J. George, Esq.

M. pichleri, Oppel, J. Carter, Q. J. G. S. XLII (1886), p. 552. *Oxford Clay; St Ives.* Presented by T. J. George, Esq.

Mecochirus pearcei, F. McCoy, A. M. N. H. ser. 2, IV (1849), p. 171, and Contrib. Brit. Pal. (1854), p. 126. J. Carter, Q. J. G. S. XLII (1866), pl. xvi, f. 13, p. 557. *Oxford Clay; Chippenham.*

Meyeria magna, F. McCoy, A. M. N. H. ser. 2, IV (1849), p. 334, and Contrib. Brit. Pal. (1854), p. 139. *Lower Greensand; Atherfield, Isle of Wight.*

M. vectensis, T. Bell, Foss. Malacos. Crustac. part ii (1862), pl. x, f. 1, p. 33. *Lower Greensand; Atherfield.* Leckenby Collection.

M. vectensis, v. Meyeria magna.

Necrocarcinus woodwardi, T. Bell, Foss. Malacos. Crustac. part ii (1862), pl. iv, f. 1, p. 20. *Chalk Marl; St Lawrence, Isle of Wight.* Leckenby Collection.

Orithopsis bonneyi, J. Carter, G. M. IX (1872), pl. xiii, f. 1, p. 531. *Upper Greensand; Lyme Regis.*

Pagurus platycheles, v. Goniochirus platycheles.

Pagurus sp., J. Carter, Q. J. G. S. XLII (1886), pl. xvi, f. 11, p. 556. *Oxford Clay; St Ives.* Presented by T. J. George, Esq.

Palæocorystes normani, T. Bell, Foss. Malacos. Crustac. part ii (1862), pl. iii, f. 10—12, p. 16. *Chalk Marl; Ventnor.* Leckenby Collection.

Palæocorystes stokesi (Mantell). *Notopocorystes mantelli,* F.
McCoy, A. M. N. H. ser. 2, IV (1849), p. 170, and Contrib.
Brit. Pal. (1854), p. 125. *Gault; Folkestone.*

Podopilumnus fittoni, F. McCoy, A. M. N. H. ser. 2, IV (1849),
p. 166, and Contrib. Brit. Pal. (1854), p. 120, woodcut. *Upper
Greensand; Lyme Regis.*

Pseudastacus? **serialis,** J. Carter, Q. J. G. S. XLII (1886), pl. xvi,
f. 12, 12 a, p. 556. *Oxford Clay; St Ives.* Presented by
A. N. Leeds, Esq.

P. sp., J. Carter, Q. J. G. S. XLII (1886), pl. xvi, f. 10, p. 556.
Oxford Clay; St Ives. Presented by T. J. George, Esq., F.G.S.

Xanthopsis bispinosa, F. McCoy, A. M. N. H. ser. 2, IV (1849),
p. 164, and Contrib. Brit. Pal. (1854), p. 119. *London Clay;
Sheppey.*

X. leachi (Desmarest). *X. nodosa,* F. McCoy, A. M. N. H. ser.
2, IV (1849), p. 163, and Contrib. Brit. Pal. (1854), p. 118.
London Clay; Sheppey.

X. unispinosa, F. McCoy, A. M. N. H. ser. 2, IV (1849), p. 164,
and Contrib. Brit. Pal. (1854), p. 119. *London Clay; Sheppey.*

VERTEBRATA.

PISCES.

Acondylacanthus attenuatus, J. W. Davis, Trans. Roy. Dublin
Soc. ser. 2, I (1883), pl. xlvi, f. 3, p. 352. *Carboniferous Limestone; Armagh.* [Spine.]

A. distans (McCoy). *Ctenacanthus distans,* F. McCoy, A. M. N.
H. ser. 2, II (1848), p. 116, and Contrib. Brit. Pal. (1854),
p. 11, and Brit. Palæoz. Foss. (1855), pl. iii K, f. 15, 15 a,
p. 625. *Carboniferous Limestone; Armagh.* [Spine.]

A. jenkinsoni (McCoy), J. W. Davis, Trans. Roy. Dublin Soc.
ser. 2, I (1883), pl. xlvi, f. 2, 2 a, p. 351. *Leptacanthus
jenkinsoni,* F. McCoy, Brit. Palæoz. Foss. (1855), pl. iii G,
f. 14—16, p. 633. *Carboniferous Limestone; Lowick, Northum-
berland.* [Spine.] The specimen figured by Davis is missing.

Acondylacanthus junceus (McCoy), J. W. Davis, Trans. Roy. Dublin Soc. ser. 2, I (1883), pl. xlvi, f. 6, 6 *a*, p. 350. *Leptacanthus junceus*, F. McCoy, A. M. N. H. ser. 2, II (1848), p. 122, and Contrib. Brit. Pal. (1854), p. 17, and Brit. Palæoz. Foss. (1855), iii G, f. 13, 13 *a*, p. 633. *Carboniferous Limestone; Derbyshire.* [Spine.]

Acrolepis hopkinsi (McCoy), F. McCoy, Brit. Palæoz. Foss. (1855), pl. iii G, f. 10, p. 609. *Holoptychius hopkinsi*, F. McCoy, A. M. N. H. ser. 2, II (1848), p. 2, and Contrib. Brit. Pal. (1854), p. 2. *Carboniferous Limestone; Derbyshire.* [Scales.]

Aspidorhynchus ornatissimus, L. Agassiz, Poiss. Foss. II (1842), pl. xlvii, p. 138. *Lithographic Slate; Solenhofen.* [Part of fish. Counterpart of the Type.] Münster Collection.

Asteroptychius ornatus, Agassiz, F. McCoy, Brit. Palæoz. Foss. (1855), pl. iii K, f. 23, 24, p. 615. *Asteroptychius semiornatus*, F. McCoy, A. M. N. H. ser. 2, II (1848), p. 118, and Contrib. Brit. Pal. (1854), p. 13, and Brit. Palæoz. Foss. (1855), pl. iii K, f. 22, 22 *a*, p. 616. *Carboniferous Limestone; Armagh.* [Spines.]

A. semiornatus, v. Asteroptychius ornatus.

Belonostomus sphyrænoides, L. Agassiz, Poiss. Foss. II (1844), pl. xlvii *a*, f. 5, p. 140. *Lithographic Slate; Eichstädt.* [Fish. Counterpart of the Type.] Münster Collection.

Centrodus striatulus, v. Megalichthys hibberti.

Cheiracanthus grandispinus, F. McCoy, A. M. N. H. ser. 2, II (1848), p. 300, and Contrib. Brit. Pal. (1854), p. 32, and Brit. Palæoz. Foss. (1855), pl. ii B, f. 1, 1 *a*, p. 582. *Old Red Sandstone; Caithness.* [Imperfect fish.] Presented by Rev. J. H. Pollexfen.

C. lateralis, v. Cheiracanthus murchisoni.

C. murchisoni, Agassiz. *C. lateralis*, F. McCoy, A. M. N. H. ser. 2, II (1848), p. 300, and Contrib. Brit. Pal. (1854), p. 32, and Brit. Palæoz. Foss. (1855), p. 582. *Old Red Sandstone; Orkney.* [Fish.]

C. pulverulentus, F. McCoy, A. M. N. H. ser. 2, II (1848), p. 299, and Contrib. Brit. Pal. (1854), p. 31, and Brit. Palæoz.

158   PISCES.

Foss. (1855), pl. ii B, f. 2, 2 a, p. 583. *Old Red Sandstone;*
Orkney. [Fish.]   Presented by Rev. J. H. Pollexfen.

Cheiracanthus pulverulentus, v. C. murchisoni.

Cheirodus pes-ranæ, F. McCoy, A. M. N. H. ser. 2, II (1848),
p. 131, and Contrib. Brit. Pal. (1854), p. 26, and Brit. Palæoz.
Foss. (1855), pl. iii G, f. 9, 9 a, p. 616. R. H. Traquair, A. M.
N. H. ser. 5, II (1878), p. 16.   J. W. Davis, Trans. Roy. Dublin
Soc. ser. 2, I (1883), pl. lxiii, f. 5, 5 a, p. 523. *Carboniferous
Limestone; Derbyshire.* [Splenial bone.]

Cheirolepis curtus, v. C. trailli.

C. macrocephalus, v. C. trailli.

Cheirolepis trailli, Agassiz. *C. curtus,* F. McCoy, A. M. N. H.
ser. 2, II (1848), p. 302, and Contrib. Brit. Pal. (1854), p. 34,
and Brit. Palæoz. Foss. (1855), pl. ii D, f. 1, 1 a, b, p. 580.
Old Red Sandstone; Lethenbar. [Fish.]

   C. macrocephalus, F. McCoy, A. M. N. H. ser. 2, II (1848),
p. 303, and Contrib. Brit. Pal. (1854), p. 35, and Brit. Palæoz.
Foss. (1855), pl. ii D, f. 3, 3 a, p. 580. *Old Red Sandstone;
Orkney.* [Fish.]

   C. velox, F. McCoy, A. M. N. H. ser. 2, II (1848), p. 302, and
Contrib. Brit. Pal. (1854), p. 34, and Brit. Palæoz. Foss. (1855),
pl. ii D, f. 2, 2 a, p. 581. *Old Red Sandstone; Orkney.* [Fish.]

C. velox, v. Cheirolepis trailli.

Chomatodus clavatus, v. Janassa clavata.

C. (Helodus) denticulatus, v. Venustodus denticulatus.

C. obliquus, v. Helodus obliquus.

C. truncatus, v. Janassa clavata.

Cladodus lœvis, v. C. marginatus.

Cladodus marginatus, Agassiz. *Cladodus lœvis,* F. McCoy,
A. M. N. H. ser. 2, II (1848), p. 133, and Contrib. Brit. Pal.
(1854), p. 28, and Brit. Palæoz. Foss. (1855), pl. iii K, f. 5,
p. 619. *Carboniferous Limestone; Armagh.* [Tooth.]

Climaxodus imbricatus, v. Janassa imbricata.

Coccoderma substriolatum (Huxley). *Macropoma substriolatum,* T. H. Huxley, Organic Remains, dec. XII (Mem. Geol.

Survey, 1866), pls. ix, x, p. 39. *Kimeridge Clay: Cottenham.* [Skull and portion of trunk.]

Coccosteus decipiens, Agassiz. *C. microspondylus*, F. McCoy, A. M. N. H. ser. 2, ii (1848), p. 298, and Contrib. Brit. Pal. (1854), p. 30, and Brit. Palæoz. Foss. (1855), pl. ii C, f. 4, p. 602. *Old Red Sandstone; Orkney.* [Imperfect skeleton.]

C. pusillus, F. McCoy, A. M. N. H. ser. 2, ii (1848), p. 298, and Contrib. Brit. Pal. (1854), p. 30, and Brit. Palæoz. Foss. (1855), pl. ii C, f. 5, 5 *a*, p. 603. *Old Red Sandstone; Orkney.* [Imperfect skeleton.] Presented by the Rev. J. H. Pollexfen.

C.? trigonaspis, F. McCoy, A. M. N. H. ser. 2, ii (1848), p. 299, and Contrib. Brit. Pal. (1854), p. 31, and Brit. Palæoz. Foss. (1855), pl. ii C, f. 6, 6 *a*, p. 603. *Old Red Sandstone; locality unknown.* [Imperfect skeleton.]

C. microspondylus, v. Coccosteus decipiens.

C. pusillus, v. Coccosteus decipiens.

C.? trigonaspis, v. Coccosteus decipiens.

Cochliodus acutus, v. Deltoptychius acutus.

C. compactus, v. Deltoptychius acutus.

Cochliodus contortus, Agassiz, F. McCoy, Brit. Palæoz. Foss. (1855), p. 622. R. Owen, G. M. iv (1867), pl. iii, f. 1, 2. J. W. Davis, Trans. Roy. Dublin Soc. ser. 2, i (1883), pl. lii, f. 6, p. 421. *Tomodus convexus*, R. Owen (*errore*), G. M. iv (1867), pl. iv, f. 2—5, p. 62. *Carboniferous Limestone; Armagh.* [Jaw.]

C. magnus, v. Psephodus magnus.

C. oblongus, v. Streblodus oblongus.

C. striatus, v. Xystrodus striatus.

Conchodus ostreæformis, F. McCoy, A. M. N. H. ser. 2, ii (1848), p. 312, and Contrib. Brit. Pal. (1854), p. 44, and Brit. Palæoz. Foss. (1855), pl. ii C, f. 7, p. 593. *Old Red Sandstone; Scat Craig, Elgin.* [Tooth.]

Ctenacanthus crenatus, F. McCoy, Brit. Palæoz. Foss. (1855), pl. iii I, f. 31, 31 *a, b*, p. 624. *C. crenulatus*, Agassiz, J. W. Davis, Trans. Roy. Dublin Soc. ser. 2, i (1883), pl. xlv, f. 6, 6 *a*, p. 345. *Carboniferous Limestone; Armagh.* [Spine.]

C. crenulatus, v. Ctenacanthus crenatus.

Ctenacanthus denticulatus, F. McCoy, A. M. N. H. ser. 2, II
(1848), p. 116, and Contrib. Brit. Pal. (1854), p. 11, and Brit.
Palæoz. Foss. (1855), pl. iii K, f. 16, p. 625. J. W. Davis, Trans.
Roy. Dublin Soc. ser. 2, I (1883), pl. xliv, f. 4, 4 a, p. 338. *Lower
Carboniferous; Monaduff*. [Spine.]

C. distans, v. Acondylacanthus distans.

C. heterogyrus, Agassiz, F. McCoy, Brit. Palæoz. Foss. (1855),
pl. iii I, f. 32, (33 missing), p. 625. *Carboniferous Limestone;
Armagh*. [Spine.]

Ctenoptychius serratus (Owen), F. McCoy, Brit. Palæoz. Foss.
(1855), pl. iii I, f. 21—23, p. 626. *Carboniferous Limestone;
f. 21, 22 from Armagh, f. 23 from Derbyshire*. [Teeth.]

Deltodus aliformis (McCoy), J. W. Davis, Trans. Roy. Dublin
Soc. ser. 2, I (1883), pl. liii, f. 12, p. 431. *Pœcilodus aliformis*,
F. McCoy, A. M. N. H. ser. 2, II (1848), p. 129, and Contrib.
Brit. Pal. (1854), p. 24, and Brit. Palæoz. Foss. (1855), pl. iii G,
f. 10 (upper figure), p. 638. *Carboniferous Limestone; Derby-
shire*. [Tooth.]

D. sublævis (Agassiz). *Pœcilodus sublœvis*, F. McCoy, Brit. Palæoz.
Foss. (1855), pl. iii I, f. 7—9, p. 640. *Pœcilodus parallelus*, F.
McCoy, *ibid*. (1855), pl. iii I, f. 6, 6 a, p. 640. *Carboniferous
Limestone; Armagh*. [Teeth.]

Deltoptychius acutus (Agassiz). *Cochliodus acutus*, F. McCoy,
Brit. Palæoz. Foss. (1855), pl. iii I, f. 24, 24 a, p. 621 (missing).
Cochliodus compactus, R. Owen, G. M. IV (1867), pl. iv, f. 1.
Carboniferous Limestone; Armagh. [Teeth.]

Dicrenodus dentatus (McCoy). *Pristicladodus dentatus*, F.
McCoy, Brit. Palæoz. Foss. (1855), pl. iii G, f. 2, 3, p. 642.
Carboniferous Limestone; Derbyshire. [Tooth.]

D. goughi (McCoy). *Pristicladodus goughi*, F. McCoy, Brit.
Palæoz. Foss. (1855), pl. iii K, f. 11, p. 643. J. W. Davis,
Trans. Roy. Dublin Soc. ser. 2, I (1883), pl. xlix, f. 27, p. 385.
Carboniferous Limestone; Kettlewell, Kendal. [Tooth.]

Diplacanthus gibbus, v. D. striatus.

Diplacanthus longispinus, Agassiz. *D. perarmatus*, F. McCoy,
A. M. N. H. ser. 2, II (1848), p. 301, and Contrib. Brit. Pal.

(1854), p. 33, and Brit. Palæoz. Foss. (1855), pl. ii B, f. 3, 3 a,
p. 585. *Old Red Sandstone; Orkney.* [Fish.]

Diplacanthus perarmatus, v. D. longispinus.

Diplacanthus striatus, Agassiz. *D. gibbus,* F. McCoy, A. M. N. H.
ser. 2, II (1848), p. 301, and Contrib. Brit. Pal. (1854), p. 33,
and Brit. Palæoz. Foss. (1855), pl. ii B, f. 4, 4 a, p. 584. *Old
Red Sandstone; Orkney.* [Fish.] Presented by Rev. J. H.
Pollexfen.

Diplopterus agassizi, Traill. *D. gracilis,* F. McCoy, A. M. N. H.
ser. 2, II (1848), p. 305, and Contrib. Brit. Pal. (1854), p. 37,
and (*Diplopterax*) Brit. Palæoz. Foss. (1855), pl. ii C, f. 1, 1 a,
b, p. 586. *Old Red Sandstone; Orkney.* [Fish.]

  Gyroptychius diplopteroides, F. McCoy, A. M. N. H. ser. 2, II
(1848), p. 309, and Contrib. Brit. Pal. (1854), p. 37, and Brit.
Palæoz. Foss. (1855), pl. ii C, f. 3, 3 a, p. 597. *Old Red
Sandstone; Orkney.* [Fish.]

D. gracilis, v. Diplopterus agassizi.

Dipriacanthus stokesi, F. McCoy, A. M. N. H. ser. 2, II (1848),
p. 121, and Contrib. Brit. Pal. (1854), p. 16, and Brit. Palæoz.
Foss. (1855), pl. iii K, f. 18, 18 a, p. 627. J. W. Davis, Trans.
Roy. Dublin Soc. ser. 2, I (1883), pl. xlviii, f. 10, 10 a, p. 360.
Carboniferous Limestone; Armagh. [Spine.]

Edaphodon laminosus, E. T. Newton, Chimæroid Fishes Brit.
Cret. Rocks (Mem. Geol. Surv. 1878), pl. viii, f. 4, 5, p. 24.
Cambridge Greensand; Cambridge. [Maxilla.]

E. reedi, E. T. Newton, Chimæroid Fishes, Brit. Cret. Rocks
(Mem. Geol. Surv. 1878), pl. vi, f. 3, 4, p. 19. *Cambridge
Greensand; Cambridge.* [Maxilla.]

E. sedgwicki (Agassiz), E. T. Newton, Chimæroid Fishes, Brit.
Cret. Rocks (Mem. Geol. Surv. 1878), pl. i, f. 9, 10, pl. ii, f. 11,
12, p. 7. *Cambridge Greensand; Cambridge.* [Maxillæ.]

  Edaphodus huxleyi, H. G. Seeley, A. M. N. H. ser. 3, XIV
(1864), p. 276. *Red Chalk; Hunstanton.* [Portions of teeth.]

Edaphodus huxleyi, v. Edaphodon sedgwicki.

Erismacanthus jonesi, F. McCoy, A. M. N. H. ser. 2, II (1848),
p. 119, and Contrib. Brit. Pal. p. 14, and Brit. Palæoz. Foss.

W. C.   11

(1855), pl. iii K, f. 26, 27, p. 628. *Carboniferous Limestone;*
Armagh. [Spine.] Presented by the Rev. W. Stokes.

Eurycormus grandis, A. Smith Woodward, Rep. Brit. Assoc.
(1889), p. 585, and G. M. dec. 3, VI (1889), p. 448, and Q. J.
G. S. XLVI (1890), p. 8 (Proc.), and G. M. dec. 3, VII (1890),
pl. x, f. 1—8, p. 289. *Kimeridge Clay; Ely.* [Head.]

Glossodus lingua-bovis, McCoy. *Glossodus marginatus,* F.
McCoy, A. M. N. H. ser. 2, II (1848), p. 128, and Contrib. Brit·
Pal. (1854), p. 23, and Brit. Palæoz. Foss. (1855), pl. iii K, f. 1,
p. 629. *Carboniferous Limestone; Armagh.* [Tooth.]

G. marginatus, v. Glossodus lingua-bovis.

Glyptolepis leptopterus, Agassiz. *Holoptychius sedgwicki,* F.
McCoy, A. M. N. H. ser. 2, II (1848), p. 311, and Contrib. Brit.
Pal. (1854), p. 43, and Brit. Palæoz. Foss. (1855), pl. ii D, f. 6,
6 a, p. 595. *Old Red Sandstone; Orkney.* [Fish.]

Gyracanthus obliquus, F. McCoy, A. M. N. H. ser. 2, II (1848),
p. 117, and Contrib. Brit. Pal. (1854), p. 12, and Brit. Palæoz.
Foss. (1855), pl. iii K, f. 13, 14, p. 629. *Lower Carboniferous;*
Moyheeland. [Spine.]

Gyroptychius angustus, v. G. microlepidotus.

G. diplopteroides, v. Diplopterus agassizi.

Gyroptychius microlepidotus (Agassiz). *G. angustus,* F. McCoy,
A. M. N. H. ser. 2, II (1848), p. 308, and Contrib. Brit. Pal.
(1854), p. 40, and Brit. Palæoz. Foss. (1855), pl. ii C, f. 2, 2 a,
p. 596. *Old Red Sandstone; Orkney.* [Imperfect fish.] Presented by Rev. J. H. Pollexfen.

Helodus didymus, v. Psephodus magnus.

Helodus mammillaris, Agassiz, F. McCoy, Brit. Palæoz. Foss.
(1855), pl. iii I, f. 16, 16 a, p. 631. *Carboniferous Limestone;*
Armagh. [Tooth.]

H. obliquus (McCoy). *Chomatodus obliquus,* F. McCoy, A. M.
N. H. ser. 2, II (1848), p. 124, and Contrib. Brit. Pal. (1854),
p. 19, and (*Helodus*) Brit. Palæoz. Foss. (1855), pl. iii K, f. 3,
p. 618. *Carboniferous Limestone; Armagh.* [Tooth.]

H. planus, v. Psephodus magnus.

Helodus pusillus (McCoy). *Polyrhizodus pusillus,* F. McCoy, A.
M. N. H. ser. 2, II (1848), p. 126 and Contrib. Brit. Pal. (1854),
p. 21, and Brit. Palæoz. Foss. (1855), pl. iii K, f. 2, 2 *a, b,*
p. 642. *Carboniferous Limestone; Armagh.* [Tooth.]

H. rudis, v. Psephodus magnus.

Holoptychius giganteus, Agassiz. *H. princeps,* F. McCoy, A.
M. N. H. ser. 2, II (1848), p. 310, and Contrib. Brit. Pal.
(1854), p. 42, and Brit. Palæoz. Foss. (1855), p. 595. *Old Red
Sandstone; Scat Craig.* [Scale.]

H. hopkinsi, v. Acrolepis hopkinsi.

H. princeps, v. Holoptychius giganteus.

H. sedgwicki, v. Glyptolepis leptopterus.

Homacanthus macrodus, F. McCoy, A. M. N. H. ser. 2, II
(1848), p. 115, and Contrib. Brit. Pal. (1854), p. 10, and Brit.
Palæoz. Foss. (1855), pl. iii K, f. 20, 20 *a,* p. 632. J. W. Davis,
Trans. Roy. Dublin Soc. ser. 2, I (1883), p. 362 (not figured).
Carboniferous Limestone; Armagh. [Spine.]

H. microdus, F. McCoy, A. M. N. H. ser. 2, II (1848), p. 115, and
Contrib. Brit. Pal. (1855), p. 10, and Brit. Palæoz. Foss. (1855),
pl. iii K, f. 19, 19 *a,* p. 633. *Carboniferous Limestone; Armagh.*
[Spine.] This specimen cannot be found.

Ischyodus brevirostris, v. I. thurmanni.

Ischyodus thurmanni, Pictet and Campiche. *I. brevirostris,*
Egerton, **var.** 2, E. T. Newton, Chimæroid Fishes Brit. Cret.
Rocks (Mem. Geol. Survey, 1878), pl. ix, f. 11, p. 31. *Lower
Chalk; Isleham, near Newmarket.* [Right mandibular ramus.]

Janassa clavata (McCoy). *Chomatodus clavatus,* F. McCoy,
Brit. Palæoz. Foss. (1855), pl. iii K, f. 10, 10 *a, b,* p. 617.
Carboniferous Limestone; Armagh. [Tooth.]
Chomatodus (Petalodus) truncatus, Agassiz, F. McCoy, *ibid.*
(1855), pl. iii I, f. 1, 1 *a,* p. 618. *Carboniferous Limestone;
Armagh.* [Tooth.]

J. imbricata (McCoy). *Climaxodus imbricatus,* F. McCoy, A. M.
N. H. ser. 2, II (1848), p. 129, and Contrib. Brit. Pal. (1854),
p. 24, and Brit. Palæoz. Foss. (1855), pl. iii G, f. 5, 5 *a,* p. 620.
Carboniferous Limestone; Derbyshire. [Tooth.]

Lepidotus maximus, Wagner. *Sphærodus neocomiensis*, Agassiz, W. Keeping, Foss. Upware and Brickhill (1883), pl. i, f. 4 *a—c*, p. 81. *Lower Greensand; Potton.* [Tooth.]

Leptacanthus jenkinsoni, v. Acondylacanthus jenkinsoni.

L. junceus, v. Acondylacanthus junceus.

Macropoma substriolatum, v. Coccoderma substriolatum.

Megalichthys hibberti, Agassiz. *Centrodus striatulus*, F. McCoy, A. M. N. H. ser. 2, II (1848), p. 4, and Contrib. Brit. Pal. (1854), p. 3, and Brit. Palæoz. Foss. (1855), pl. iii G, f. 1, 1 *a—d*, p. 611. *Carboniferous Shale; Carluke.* [Tooth.]

Osteolepis brevis, v. O. macrolepidotus.

Osteolepis macrolepidotus, Agassiz. *O. brevis*, F. McCoy, A. M. N. H. ser. 2, II (1848), p. 305, and Contrib. Brit. Pal. (1854), p. 37, and Brit. Palæoz. Foss. (1855), pl. ii D, f. 4, 4 *a*, p. 587. *Old Red Sandstone.* Labelled *Caithness*, but in a note on the back of the specimen Dr Traquair says 'I believe this to be from Cromarty—it is certainly not from Orkney or Caithness.' [Crushed fish.]

Tripterus pollexfeni, F. McCoy, A. M. N. H. ser. 2, II (1848), p. 306, and Contrib. Brit. Pal. (1854), p. 38, and (*Triplopterus*) Brit. Palæoz. Foss. (1855), pl. ii D, f. 5, 5 *a*, *b*, p. 589. *Old Red Sandstone; Orkney.* [Portion of fish.] Presented by the Rev. J. H. Pollexfen.

Osteoplax erosa, F. McCoy, A. M. N. H. ser. 2, II (1848), p. 6, and Contrib. Brit. Pal. (1854), p. 6, and Brit. Palæoz. Foss. (1855), pl. iii K, f. 12, 12 *a*, p. 613. *Carboniferous Series; Cultra, co. Down.* [Dermal plate.]

Pachyrhizodus subulidens (Owen), A. S. Woodward, A. M. N. H. dec. 6, IV (1889), p. 350. *Rhaphiosaurus subulidens*, R. Owen, Rep. Brit. Assoc. (1841), p. 190, and Rept. Cret. Form. (1851), pl. x, f. 5, 6, p. 19. *Lower Chalk; Cherry Hinton.* [Mandibular ramus.] Presented by J. Carter, Esq., F.G.S.

Petalodus acuminatus, Agassiz, F. McCoy, Brit. Palæoz. Foss. (1855), pl. iii G, f. 4, 4 *c*, p. 635. *Petalodus rhombus*, F. McCoy, A. M. N. H. ser. 2, II (1848), p. 125, and Contrib. Brit. Pal. (1854), p. 20. *Carboniferous Limestone; Derbyshire.* [Tooth.]

Petalodus psittacinus, v. Petalorhynchus psittacinus.

P. sagittatus, v. Petalorhynchus psittacinus.

P. rhombus, v. Petalodus acuminatus.

Petalorhynchus psittacinus (Agassiz). *Petalodus psittacinus*, F. McCoy, Brit. Palæoz. Foss. (1855), pl. iii I, f. 4, p. 636. *Carboniferous Limestone ; Armagh.* [Tooth.]
Petalodus sagittatus, Agassiz, F. McCoy, Brit. Palæoz. Foss. (1855), pl. iii I, f. 2, 3, 3 *a—c*, p. 636. *Carboniferous Limestone ; Armagh.* [Tooth.]

Petrodus patelliformis, F. McCoy, A. M. N. H. ser. 2, II (1848), p. 132, and Contrib. Brit. Pal. (1854), p. 27, and Brit. Palæoz. Foss. (1855), pl. iii G, f. 6—8, p. 637. J. W. Davis, Trans. Roy. Dublin Soc. ser. 2, I (1883), pl. li, f. 16, 16 *a, b,* p. 400. *Carboniferous Limestone ; Derbyshire.* [Dermal tubercles.]

Pholidophorus latimanus, L. Agassiz, Poiss. Foss. II (1838), pl. xliii, f. 2, p. 280. *Lithographic Slate ; Solenhofen.* [Part of fish. Counterpart of the Type.] Münster Collection.

P. micronyx, L. Agassiz, Poiss. Foss. II (1838), pl. xlii, f. 1, p. 279. *Lithographic Slate ; Kelheim.* [Fish.] Münster Collection.

P. tenuiserratus [Münster MS.], L. Agassiz, Poiss. Foss. II (1838), pl. xlii, f. 4, p. 276. *Lithographic Slate ; Kelheim.* [Fish.] Münster Collection.

Physonemus arcuatus, F. McCoy, A. M. N. H. ser. 2, II (1848), p. 117, and Contrib. Brit. Pal. (1854), p. 12, and Brit. Palæoz. Foss. (1855), pl. iii I, f. 29, 29 *a, b,* p. 638. *Carboniferous Limestone ; Armagh.* [Spine.]

Pœcilodus aliformis, v. Deltodus aliformis.

Pœcilodus foveolatus, F. McCoy, A. M. N. H. ser. 2, II (1848), p. 129, and Contrib. Brit. Pal. (1854), p. 24, and Brit. Palæoz. Foss. (1855), pl. iii G, f. 11, p. 639. J. W. Davis, Trans. Roy. Dublin Soc. ser. 2, I (1883), pl. liii, f. 26, p. 445. *Carboniferous Limestone ; Derbyshire.* [Tooth.]

P. jonesi, Agassiz. *P. obliquus*, Agassiz, F. McCoy, Brit. Palæoz. Foss. (1855), pl. iii I, f. 5, 5 *a,* p. 640. *Carboniferous Limestone ; Armagh.* [Teeth.]

P. obliquus, v. Pœcilodus jonesi.

Pœcilodus parallelus, v. Deltodus sublævis.

P. sublævis, v. Deltodus sublævis.

Polyrhizodus magnus, F. McCoy, A. M. N. H. ser. 2, ii (1848), p. 126, and Contrib. Brit. Pal. (1854), p. 21, and Brit. Palæoz. Foss. (1855), pl. iii κ, f. 6—8, p. 641. *Carboniferous Limestone; Armagh.* [Tooth.]

P. pusillus, v. Helodus pusillus.

Prionolepis angustus, P. M. G. Egerton, in F. Dixon, Geol. Sussex, (1850), pl. xxxii*, f. 3. *Lower Chalk; Burwell.* [Scales. Counterpart of the Type.]

Pristicladodus dentatus, v. Dicrenodus dentatus.

P. goughi, v. Dicrenodus goughi.

Psammodus canaliculatus, v. P. rugosus.

Psammodus rugosus, Agassiz. *P. canaliculatus*, F. McCoy, A. M. N. H. ser. 2, ii (1848), p. 122, and Contrib. Brit. Pal. (1854), p. 17, and Brit. Palæoz. Foss. (1855), pl. iii G, f. 12, p. 643. *Carboniferous Limestone; Armagh.* [Tooth.]

Psephodus magnus (Agassiz). *Cochliodus magnus*, F. McCoy, Brit. Palæoz. Foss. (1855), p. 622. *Helodus planus*, F. McCoy, *ibid.* (1855), pl. iii I, f. 12—15, p. 631. *Helodus didymus*, F. McCoy, *ibid.* (1855), pl. iii I, f. 18—20, p. 630 (*pars*). *Helodus rudis*, F. McCoy, A. M. N. H. ser. 2, ii (1848), p. 123, and Contrib. Brit. Pal. (1854), p. 18, and Brit. Palæoz. Foss. (1855), pl. iii κ, f. 4, 4 *a*, p. 631. *Carboniferous Limestone; Armagh.* [Teeth.]

Rhizodus gracilis, v. R. hibberti.

Rhizodus hibberti (Agassiz and Hibbert). *Rhizodus gracilis*, F. McCoy, Brit. Palæoz. Foss. (1855), pl. iii G, f. 17, p. 611. *Lower Carboniferous; Gilmerton.* [Dentary.]
Another specimen, L. C. Miall, Q. J. G. S. xxxi (1875), p. 624, woodcut. *Lower Carboniferous; Gilmerton, near Edinburgh.* [Skull.]

Scaphodus heteromorphus, A. Smith Woodward, Proc. Geol. Assoc. xi (1889), pl. iii, f. 33, p. 300. *Stonesfield Slate.* [Lower dentition.]

Sphærodus neocomiensis, v. Lepidotus maximus.

Steganodictyum carteri, F. McCoy, A. M. N. H. ser. 2, VIII (1851), p. 483, and Contrib. Brit. Pal. (1854), p. 234, and Brit. Palæoz. Foss. (1855), pl. ii A, f. 4, 4 *a*. *Devonian; Llantivet Bay, Cornwall.* [Fragment of shield.]

S. cornubicum, F. McCoy, A. M. N. H. ser. 2, VIII (1851), p. 482, Contrib. Brit. Pal. (1854), p. 233, and Brit. Palæoz. Foss. (1855), pl. ii A, f. 1—3. *Devonian; Llantivet Bay, Cornwall.* [Fragments of shields.]

Streblodus oblongus (Portlock). *Cochliodus oblongus,* F. McCoy, Brit. Palæoz. Foss. (1855), pl. iii H, f. 19, pl. iii I, f. 28, p. 623. *Carboniferous Limestone; Armagh.* [Teeth.]

Thrissopater salmoneus, A. C. L. G. Günther, Brit. Organic Remains, dec. XIII (Mem. Geol. Survey, 1872), no. 1, pl. i. *Gault; East Wear Bay, Folkestone.* [Fish.] Presented by A. S. Reid, Esq.

Thyellina angusta [Münster MS.], L. Agassiz, Poiss. Foss. III (1838), pl. xxxix, f. 3, p. 378. *Chalk; Baumberg.* [Fish.] Münster Collection.

Triplopterus pollexfeni, v. Osteolepis macrolepidotus.

Tripterus pollexfeni, v. Osteolepis macrolepidotus.

Venustodus denticulatus (McCoy). *Chomatodus* (*Helodus*) *denticulatus,* F. McCoy, A. M. N. H. ser. 2, II (1848), p. 124, and Contrib. Brit. Pal. (1854), p. 19, and Brit. Palæoz. Foss. (1855), pl. iii K, f. 9, 9 *a*, p. 618. *Carboniferous Limestone; Armagh.* [Tooth.]

Xystrodus striatus (Agassiz). *Cochliodus striatus,* F. McCoy, Brit. Palæoz. Foss. (1855), pl. iii I, f. 27, 27 *a*, p. 624. *Carboniferous Limestone; Armagh.* [Tooth.]

AMPHIBIA.

Keraterpeton?, Rep. Rugby School Nat. Hist. Soc. (1874), pl. i, p. 52. T. Oldham, *ibid.* (1875), pl. vii, p. 74. J. M. Wilson, Q. J. G. S. XXXI (1875), p. 81 (Proc.). *Coal Measures; near Castlecomer, Ireland.* [Impression of skull and vertebral column.]

168   REPTILIA.

REPTILIA.

Acanthopholis eucercus, H. G. Seeley, Q. J. G. S. xxxv (1879),
p. 632. *Cambridge Greensand; Cambridge.* [Vertebræ.]

A. macrocercus, H. G. Seeley, Index to Aves, etc., Woodw. Mus.
(1869), p. xvii. *Cambridge Greensand; Cambridge.* [Ver-
tebræ.]

A.? platypus, H. G. Seeley, Index to Aves, etc., Woodw. Mus.
(1869), p. xvii, and A. M. N. H. ser. 4, VIII (1871), pl. vii,
p. 305. *Cambridge Greensand; Cambridge.* [Metapodium.]

A. stereocercus, H. G. Seeley, Index to Aves, etc., Woodw. Mus.
(1869), p. xvii, and Q. J. G. S. xxxv (1879), p. 628. *Cambridge
Greensand; Cambridge.* [Vertebræ.]

A.?, H. G. Seeley, Q. J. G. S. xxxv (1879), p. 594, woodcut 1.
Cambridge Greensand; Cambridge. [Axis vertebra.]

Anoplosaurus curtonotus, H. G. Seeley, Q. J. G. S. xxxv
(1879), pls. xxxiv, xxxv, p. 600. *Cambridge Greensand; Reach,
near Cambridge.* [Remains of skeleton.]

A. major, H. G. Seeley, Q. J. G. S. xxxv (1879), p. 631.
Cambridge Greensand; Cambridge. [Cervical vertebra.]

Chelone planimentum, v. Lytoloma planimentum.

Cimoliosaurus bernardi (Owen). *? Plesiosaurus pachyomus*, R.
Owen (*pars*), Rep. Brit. Assoc. (1839), p. 74, and Rept. Cret.
Form. (1851), pl. xx, f. 1—6, p. 64. *Cambridge Greensand;
Cambridge.* [Vertebræ.] Presented by J. Carter, Esq., F.G.S.
   Plesiosaurus ichthyospondylus, H. G. Seeley, Index to Aves,
etc., Woodw. Mus. (1869), p. xvii. *Cambridge Greensand;
Cambridge.* [Vertebræ, humerus, etc.]

C. planus (Owen). *Plesiosaurus planus*, R. Owen, Rept. Cret.
Form. Suppl. iv (1864), pl. i, f. 1—26, p. 2. *Cambridge
Greensand; Cambridge.* [Vertebræ.]
   Plesiosaurus pachyomus, R. Owen (*pars*), Rep. Brit. Assoc.
(1839), p. 74, and Rept. Cret. Form. (1851), pl. xxi, f. 1—6,
p. 64. *Cambridge Greensand; Cambridge.* [Vertebræ.] Presented by J. Carter, Esq.

C. portlandicus (Owen). *Plesiosaurus winspitensis*, H. G. Seeley,

A. M. N. H. ser. 4, VIII (1871), p. 181, woodcuts. *Portland Oolite; Winspit.* [Vertebræ.]

Cimoliosaurus trochanterius (Owen), R. Lydekker, G. M. dec. 3, V (1888), pp. 354—6. *Plesiosaurus megadeirus,* H. G. Seeley, Index to Aves, etc., Woodw. Mus. (1869), pp. xx, 97, 121. *Colymbosaurus megadeirus,* H. G. Seeley, Q. J. G. S. xxx (1874), p. 445. *Kimeridge Clay; Ely.* [Coracoids, scapula, etc.]

Coloborhynchus sedgwicki, v. Ornithocheirus sedgwicki.

Colymbosaurus megadeirus, v. Cimoliosaurus trochanterius.

Craterosaurus pottonensis, H. G. Seeley, Q. J. G. S. xxx (1874), pl. xliv, p. 690. *Lower Greensand; Potton.* [Portion of cranium.]

Criorhynchus simus, v. Ornithocheirus simus.

Crocodilus? **cantabrigiensis,** H. G. Seeley, Index to Aves, etc., Woodw. Mus. (1869), p. xvii, and Q. J. G. S. xxx (1874), p. 693, woodcuts. *Cambridge Greensand; Cambridge.* [Vertebra.]

C.? **icenicus,** H. G. Seeley, Q. J. G. S. xxxii (1876), p. 437. *Cambridge Greensand; Cambridge.* [Vertebra.]

Cryptodraco eumerus (Seeley), R. Lydekker, Q. J. G. S. xLV (1889), p. 45. *Cryptosaurus eumerus,* H. G. Seeley, Index to Aves, etc., Woodw. Mus. (1869), p. 93, and Q. J. G. S. xxxi (1875), pl. vi, p. 149. *Oxford Clay; Great Gransden.* [Femur.] Presented by L. Ewbank, Esq., M.A.

Cryptosaurus eumerus, v. Cryptodraco eumerus.

Dacosaurus maximus (Plieninger). *Dakosaurus lissocephalus,* H. G. Seeley, Index to Aves, etc., Woodw. Mus. (1869), p. 92. *Kimeridge Clay; Ely.* [Skull.]

Dakosaurus lissocephalus, v. Dacosaurus maximus.

Doratorhynchus validum (Owen), H. G. Seeley, Q. J. G. S. xxxi (1875), p. 465, woodcut. *Pterodactylus macrurus,* H. G. Seeley, Index to Aves, etc., Woodw. Mus. (1869), p. 89, and Proc. Camb. Phil. Soc. II (1869), p. 130. *Purbeck Beds; Langton, near Swanage.* [Mandible and caudal vertebra.]

Emys hordwellensis, v. Ocadia crassa.

170   REPTILIA.

Enaliochelys chelonia, v. Thalassemys hugii.

Eretmosaurus macropterus (Seeley). *Plesiosaurus macropterus*, H. G. Seeley, A. M. N. H. ser. 3, xv (1865), pp. 49, 232. *Lias; Whitby.* [Skeleton.]

Eucercosaurus tanyspondylus, H. G. Seeley, Q. J. G. S. xxxv (1879), p. 613, woodcuts 4, 5. *Cambridge Greensand; Trumpington, near Cambridge.* [Vertebræ.]

Gigantosaurus megalonyx, H. G. Seeley, Index to Aves, etc., Woodw. Mus. (1869), p. 94. *Kimeridge Clay; Ely.* [Vertebræ and detached limb-bones.]

Glossochelys harvicensis, v. Lytoloma planimentum.

Ichthyosaurus angustidens, H. G. Seeley, Index to Aves, etc., Woodw. Mus. (1869), pp. xv, 3. *Chalk; Hunstanton.* [Tooth.]

I. bonneyi, H. G. Seeley, Index to Aves, etc., Woodw. Mus. (1869), p. xvii. *Cambridge Greensand; Cambridge.* [Femur.]

Ichthyosaurus campylodon, J. Carter, Rep. Brit. Assoc. (1845), Sect. p. 60, and London Geol. Journ. I (1846), p. 7, woodcut. R. Owen, Rept. Cret. Form. (1851), pl. iv, f. 1—10, 13—16 [f. 7 is missing] and pl. xxv, f. 1, 2, pp. 72, 75. *Lower Chalk, and Cambridge Greensand; Cambridgeshire.* [Teeth.] Presented by J. Carter, Esq.

I. chalarodeirus, H. G. Seeley, Index to Aves, etc., Woodw. Mus. (1869), pp. xx, 106. *Kimeridge Clay; Ely.* [Axis vertebra.]

I. doughtyi, H. G. Seeley, Index to Aves, etc., Woodw. Mus. (1869), p. xvii. *Cambridge Greensand; Cambridge.* [Humerus.]

I. hygrodeirus, H. G. Seeley, Index to Aves, etc., Woodw. Mus. (1869), pp. xx, 106. *Kimeridge Clay; Ely.* [Vertebra.]

I. megalodeirus, H. G. Seeley, Index to Aves, etc., Woodw. Mus. (1869), pp. xxi, 110. *Oxford Clay; Woodstone Lodge, near Peterborough.* [Parts of skeleton.]

I. platymerus, H. G. Seeley, Index to Aves, etc., Woodw. Mus. (1869), p. xvii. *Cambridge Greensand; Cambridge.* [Femur.]

I. zetlandicus, H. G. Seeley, Q. J. G. S. xxxvi (1880), pl. xxv, p. 635. R. Lydekker, Cat. Foss. Rept. B. M. part ii (1889), p. 78. *Upper Lias; Whitby.* [Cranium.] Presented by Earl Zetland.

Ichthyosaurus sp. (young individual), H. G. Seeley, Rep. Brit. Assoc. (1880), pl. i, f. 1, p. 68. *Lias; Street.* [Skull and portion of vertebral column.] Presented by T. Hawkins, Esq.

Iguanodon bernissartensis, Boulenger. *? Ornithopsis?*, H. G. Seeley, Q. J. G. S. xxxviii (1882), p. 367, woodcuts 1—4. R. Lydekker, Q. J. G. S. xliv (1888), p. 52. *Wealden; Brook, Isle of Wight.* [Coracoid.]

I. phillipsi, v. Priodontognathus phillipsi.

I.? sp., H. G. Seeley, Q. J. G. S. xxxi (1875), p. 461, woodcuts. *Wealden; Brook, Isle of Wight.* [Axis.]

Lytoloma planimentum (Owen). *Chelone planimentum*, R. Owen, Rep. Brit. Assoc. (1841), p. 178, and Proc. Geol. Soc. iii (1841), p. 576. R. Owen and T. Bell, Rept. London Clay (1849), pl. ix, f. 1—3, p. 25 [the carapace figured on pl. x has perished]. *Glossochelys harvicensis* (S. Woodward), H. G. Seeley, A. M. N. H. ser. 4, viii (1871), p. 232, woodcut. *London Clay; near Harwich.* [Skull.]

Macrurosaurus semnus, H. G. Seeley, Index to Aves, etc., Woodw. Mus. (1869), pp. xvii, 45, and Q. J. G. S. xxxii (1876), p. 440, woodcuts 1, 2. *Cambridge Greensand; Cambridge.* [Caudal vertebræ.]

Metriorhynchus superciliosum (De Blainville). *? Steneosaurus dasycephalus*, H. G. Seeley, Index to Aves, etc., Woodw. Mus. (1869), pp. xxi, 140. *Oxford Clay; Peterborough.* [Skull and vertebræ.]

Ocadia crassa (Owen). *Emys hordwellensis*, H. G. Seeley, Q. J. G. S. xxxii (1876), p. 445, woodcuts 1, 2. *Lower Headon Beds; Hordwell Cliff.* [Imperfect shell.]

Ornithocheirus brachyrhinus, H. G. Seeley, Ornithosauria (1870), p. 123. *Cambridge Greensand; Cambridge.* [Fragment of premaxilla.]

O. capito, H. G. Seeley, Ornithosauria (1870), p. 126. *Cambridge Greensand; Cambridge.* [Fragment of premaxilla.]

O. carteri, H. G. Seeley, Ornithosauria (1870), pp. xvi, 6 (*Pterodactylus*). *Cambridge Greensand; Cambridge.* [Premaxilla.]

Ornithocheirus colorhinus, H. G. Seeley, Ornithosauria (1870), p. 124. *Cambridge Greensand; Cambridge.* [Fragments of premaxillæ.]

O. compressirostris (Owen), H. G. Seeley, Ornithosauria (1870), p. 92. *Pterodactylus compressirostris*, R. Owen, Rept. Cret. Form. (1851), pl. xxxii, f. 6, 6', 7, 7' (f. 8 is missing), p. 95. *Cambridge Greensand; Cambridge.* [Distal end of ulna, etc.] Presented by J. Carter, Esq., F.G.S.

O. crassidens, H. G. Seeley, Ornithosauria (1870), p. 122. *Cambridge Greensand; Cambridge.* [Fragment of ?premaxilla.]

O. cuvieri (Bowerbank), H. G. Seeley, Ornithosauria (1870), p. 113 (? pl. xii, f. 5). *Cambridge Greensand; Cambridge.* [Portion of premaxilla.]
  Pterodactylus cuvieri, R. Owen, Rept. Cret. Form. Suppl. iii (1861), pl. iii, f. 1—3. *Upper Chalk; Kent.* [Proximal end of humerus.]

O. dentatus, H. G. Seeley, Ornithosauria (1870), p. 119. *Cambridge Greensand; Cambridge.* [Fragment of premaxilla.]

O. denticulatus, H. G. Seeley, Ornithosauria (1870), pl. xii, f. 8, 9, p. 122. *Cambridge Greensand; Cambridge.* [Fragment of premaxilla.]

O. enchorhynchus, H. G. Seeley, Ornithosauria (1870), p. 123. *Cambridge Greensand; Cambridge.* [Premaxilla.]

O. eurygnathus, H. G. Seeley, Ornithosauria (1870), p. 123. *Cambridge Greensand; Cambridge.* [Fragment of ?dentary bone.]

O. fittoni (Owen), H. G. Seeley, Ornithosauria (1870), p. 118. *Pterodactylus fittoni*, R. Owen, Rep. Brit. Assoc. (1858), p. 98, and Rept. Cret. Form. Suppl. i (1859), pl. i, f. 3 (4, 5, missing), pls. ii—iv (*pars*), p. 4, and Phil. Trans. CXLIX, 1859 (1860), pl. x, f. 7—12, 32—34, p. 162. *Cambridge Greensand; Cambridge.* [Premaxilla.]

O. machærorhynchus, H. G. Seeley, Ornithosauria (1870), pl. xii, f. 1, 2, p. 113. *Cambridge Greensand; Cambridge.* [Dentary bone.]

Ornithocheirus microdon, H. G. Seeley, Ornithosauria (1870), pl. xii, f. 6, 7, p. 116. *Cambridge Greensand; Cambridge.* [Premaxilla.]

O. nasutus, H. G. Seeley, Ornithosauria (1870), p. 120. *Cambridge Greensand; Cambridge.* [Fragment of premaxilla.]

O. oweni, H. G. Seeley, Ornithosauria (1870), p. 115. *Cambridge Greensand; Cambridge.* [Premaxilla.]

O. oxyrhinus, H. G. Seeley, Ornithosauria (1870), p. 117. *Cambridge Greensand; Cambridge.* [Premaxilla.]

O. platyrhinus, H. G. Seeley, Index to Aves, etc., Woodw. Mus. (1869), p. xvii. *Cambridge Greensand; Cambridge.* [Premaxilla.]

O. platysomus, H. G. Seeley, Ornithosauria (1870), p. 120. *Cambridge Greensand; Cambridge.* [Fragment of snout.]

O. polyodon, H. G. Seeley, Ornithosauria (1870), p. 121. *Cambridge Greensand; Cambridge.* [Premaxilla.]

O. scaphorhynchus, H. G. Seeley, Ornithosauria (1870), p. 119. *Cambridge Greensand; Cambridge.* [Fragment of premaxilla.]

O. sedgwicki (Owen), H. G. Seeley, Ornithosauria (1870), p. 112. *Pterodactylus sedgwicki**, R. Owen, Rep. Brit. Assoc. (1858), p. 98, and Phil. Trans. CXLIX, 1859 (1860), pl. x, f. 1—5, 13, 14, 24—26, 28—31, 35, 36, p. 160, and Rept. Cret. Form. Suppl. i (1859), pl. i, f. 1, 2, 6—14, pls. ii—iv (*pars*), p. 2, and *ibid.* Suppl. iii (1861), pl. iv, f. 5—12. *Coloborhynchus sedgwicki,* R. Owen, Rept. Mesoz. Form. part i (1874), p. 6. *Cambridge Greensand; Cambridge.* [Portions of jaws.]

O. simus (Owen), H. G. Seeley, Index to Aves, etc., Woodw. Mus. (1869), p. xvi (*Pterodactylus,* p. 6), and Ornithosauria (1870), p. 127, and G. M. dec. 2, VIII (1881), p. 15. *Pterodactylus simus**, R. Owen, Rept. Cret. Form. Suppl. iii (1861), pls. i, ii, pl. iv, f. 4, p. 2. *Criorhynchus simus,* R. Owen, Rept. Mesoz. Form. part i (1874), p. 6. [Mandibular symphysis.]

? *Pterodactylus woodwardi*, R. Owen, Rept. Cret. Form. Suppl. iii (1861), pl. ii, f. 3, p. 4. ? *Ornithocheirus woodwardi,* H. G. Seeley, Ornithosauria (1870), p. 125. [Portion of jaw.] *Cambridge Greensand; Cambridge.*

---

* Some of the specimens figured by Owen are missing.

174 REPTILIA.

Ornithocheirus tenuirostris, H. G. Seeley, Ornithosauria (1870), p. 114. *Cambridge Greensand; Cambridge.* [Premaxilla.]

O. woodwardi, v. Ornithocheirus simus.

Ornithopsis? (Seeley), v. Iguanodon bernissartensis.

Ornithostoma sp., H. G. Seeley, A. M. N. H. ser. 4, VII (1871), p. 35 (footnote), and Q. J. G. S. XXXII (1876), p. 499. 'Proximal end of metacarpal of wing-finger of *Pterodactylus*,' R. Owen, Rept. Cret. Form. Suppl. i (1859), pl. iv, f. 4, 5, p. 18. *Cambridge Greensand; Cambridge.* [? Premaxilla.]

Patricosaurus merocratus, H. G. Seeley, Q. J. G. S. XLIII (1887), pl. xii, f. 9—12, p. 216. *Cambridge Greensand; Cambridge.* [Head of femur, and first sacral vertebra.] Presented by J. Carter, Esq.

Pelagosaurus brongniarti (Kaup). *? Teleosaurus eucephalus*, H. G. Seeley, Q. J. G. S. XXXVI (1880), pl. xxiv, p. 627. *Teleosaurus sp.*, H. G. Seeley, Index to Aves, etc., Woodw. Mus. (1869), p. 121. *Lias; Whitby.* [Imperfect cranium.]

Peloneustes æqualis (Phillips). *Plesiosaurus sterrodeirus*, H. G. Seeley, Index to Aves, etc., Woodw. Mus. (1869), pp. xx, 98. *Kimeridge Clay; Ely.* [Basi-occipital and vertebræ.]

P. philarchus (Seeley), R. Lydekker, Q. J. G. S. XLV (1889), p. 48. *Plesiosaurus philarchus*, H. G. Seeley, Index to Aves, etc., Woodw. Mus. (1869), pp. xxi, 139. *Oxford Clay; Peterborough.* [Imperfect skeleton.]

Plesiosaurus cliduchus, v. P. dolichodeirus.

P. cycnodeirus, H. G. Seeley, Index to Aves, etc., Woodw. Mus. (1869), pp. xvii, 41. *Cambridge Greensand; Cambridge.* [Vertebræ.]

P. dolichodeirus, Conybeare. *P. cliduchus*, H. G. Seeley, A. M. N. H. ser. 3, XVI (1865), pl. xv, p. 356. J. W. Hulke, Q. J. G. S. XXXIX (1883), p. 58 (Proc.), woodcut 14. *Lias; Street.* [Imperfect skeleton.] Presented by T. Hawkins, Esq.

Plesiosaurus eleutheraxon, H. G. Seeley, A. M. N. H. ser. 3, XVI (1865), pl. xiv, p. 353, and Index to Aves, etc., Woodw. Mus. (1869), pp. xxi, 137. *Plesiosaurus*, L. Barrett, A. M. N. H. ser. 3, II (1858), pl. xiii, p. 361. *Lower Lias; Street.*

[Imperfect skeleton, wanting head.] Presented by T. Hawkins, Esq.

Plesiosaurus euryspondylus, H. G. Seeley, Index to Aves, etc., Woodw. Mus. (1869), pp. xviii, 41. *Cambridge Greensand; Cambridge.* [Vertebrae.]

P. ichthyospondylus, v. Cimoliosaurus bernardi.

P. macropterus, v. Eretmosaurus macropterus.

P. megadeirus, v. Cimoliosaurus trochanterius.

P. microdeirus, H. G. Seeley, Index to Aves, etc., Woodw. Mus. (1869), pp. xviii, 44. *Cambridge Greensand; Cambridge.* [Portions of skeleton.]

P. pachyomus, v. Cimoliosaurus bernardi, and C. planus.

P. philarchus, v. Peloneustes philarchus.

P. planus, v. Cimoliosaurus planus.

P. platydeirus, H. G. Seeley, Index to Aves, etc., Woodw. Mus. (1869), pp. xviii, 41. *Cambridge Greensand; Cambridge.* [Vertebræ.]

P. pœcilospondylus, H. G. Seeley, Index to Aves, etc. Woodw. Mus. (1869), pp. xviii, 48. *Cambridge Greensand; Cambridge.* [Vertebræ.]

P. sterrodeirus, v. Peloneustes æqualis.

P. winspitensis, v. Cimoliosaurus portlandicus.

P. sp., W. Keeping, Foss. Upware and Brickhill (1883), pl. i, f. 3 *a*, *b*, p. 78. *Lower Greensand; Brickhill.* [Vertebra.]

P. sp. (? neocomiensis, Campiche), W. Keeping, Foss. Upware and Brickhill (1883), pl. i, f. 2 *a*, *b*, p. 78. *Lower Greensand; Upware.* [Vertebra.]

Pleurosternon oweni, H. G. Seeley, Index to Aves, etc., Woodw. Mus. (1869), pp. xix, 87. *Purbeck Beds; Swanage.* [Carapace.]

P. sedgwicki, H. G. Seeley, Index to Aves, etc., Woodw. Mus. (1869), pp. xix, 86. *Purbeck Beds; Swanage.* [Carapace.]

P. typocardium, H. G. Seeley, Index to Aves, etc., Woodw. Mus. (1869), pp. xix, 87. *Purbeck Beds; Swanage.* [Carapace.]

P. vansittarti, H. G. Seeley, Index to Aves, etc. Woodw. Mus. (1869), pp. xix, 86. *Purbeck Beds; Swanage.* [Carapace.]

176   REPTILIA.

Pliosaurus brachydeirus, Owen, H. G. Seeley, Index to Aves, etc., Woodw. Mus. (1869), p. 104. *Kimeridge Clay; Cottenham.* [Vertebra.] Presented by the Rev. S. Banks, M.A.

P. brachyspondylus (Owen), H. G. Seeley, Index to Aves, etc., Woodw. Mus. (1869), p. 102. *Kimeridge Clay; Ely.* [Vertebræ.]

P. evansi, H. G. Seeley, Index to Aves, etc., Woodw. Mus. (1869), pp. xxi, 116, and Q. J. G. S. xxxiii (1877), p. 716, woodcuts 1—9, and G. M. dec. 3. iv (1887), p. 478. R. Lydekker, Q. J. G. S. xlvi (1890), p. 51. *Oxford Clay; Eynsbury, near St Neots.* [Atlas and axis.]

P. ferox (Sauvage). ? *Pliosaurus pachydeirus*, H. G. Seeley, Index to Aves, etc., Woodw. Mus. (1869), pp. xxi, 118. R. Lydekker, G. M. dec. 3, v (1888), p. 353, and Q. J. G. S. xlvi (1890), p. 51. *Oxford Clay; Great Gransden.* [Cervical vertebræ.] Presented by L. Ewbank, Esq., M.A.

P. microdeirus, H. G. Seeley, Index to Aves, etc., Woodw. Mus. (1869), pp. xx, 76. *Lower Greensand; Wicken.* [Cervical vertebra.]

P. pachydeirus, v. Pliosaurus ferox.

Polyptychodon interruptus, R. Owen, Rept. Cret. Form. (1851) pl. x, f. 8, 9, p. 55 [tooth presented by J. Carter, Esq.], and *ibid.* Suppl. iii (1861), pl. v, f. 1—3, pl. vi, f. 1—4, p. 20, and Q. J. G. S. xvi (1860), p. 262. H. G. Seeley, Q. J. G. S. xxxii (1876), p. 433. *Cambridge Greensand; Cambridge.* [Tooth and vertebræ.]

Priodontognathus phillipsi (Seeley), H. G. Seeley, Q. J. G. S. xxxi (1875), pl. xx, p. 439. *Iguanodon phillipsi*, H. G. Seeley, Index to Aves, etc., Woodw. Mus. (1869), pp. xix, 82. *Locality unknown.* [Maxilla.]

Pterodactylus compressirostris, v. Ornithocheirus compressirostris.

P. cuvieri, v. Ornithocheirus cuvieri.

P. fittoni, v. Ornithocheirus fittoni.

P. hopkinsi, H. G. Seeley, Proc. Camb. Phil. Soc. i (1864), p. 228. *Cambridge Greensand; Cambridge.* [Premaxilla.]

P. macrurus, v. Doratorhynchus validum.

Pterodactylus oweni, H. G. Seeley, Proc. Camb. Phil. Soc. i (1864), p. 228. *Cambridge Greensand ; Cambridge.* [Premaxilla.]

P. sedgwicki, v. Ornithocheirus sedgwicki.

P. simus, v. Ornithocheirus simus.

P. woodwardi, v. Ornithocheirus simus.

Rhaphiosaurus subulidens, v. Pachyrhizodus subulidens.

Rhinochelys pulchriceps, H. G. Seeley, Index to Aves, etc., Woodw. Mus. (1869), pp. xviii, 25. *Cambridge Greensand ; Cambridge.* [Skull.]

Saurospondylus dissimilis, H. G. Seeley, A. M. N. H. ser. 3, xvi (1865), p. 145, and Index to Aves, etc., Woodw. Mus. (1869), pp. xv, 3. *Lower Chalk ; Cherry Hinton.* [Vertebra.]

Sphenospondylus gracilis, Lydekker. *Sphenospondylus*, H. G. Seeley, Q. J. G. S. xxxix (1883), p. 55, woodcuts 1—3. *Wealden ; Brook, Isle of Wight.* [Dorsal vertebra.]

Steneosaurus dasycephalus, v. Metriorhynchus superciliosum.

Stereosaurus cratynotus, H. G. Seeley, Index to Aves, etc., Woodw. Mus. (1869), pp. xviii, 44. *Cambridge Greensand ; Cambridge.* [Vertebræ.]

S. platysomus, H. G. Seeley, Index to Aves, etc., Woodw. Mus. (1869), pp. xviii, 43. *Cambridge Greensand ; Cambridge.* [Vertebræ.]

S. stenomus, H. G. Seeley, Index to Aves, etc., Woodw. Mus. (1869), pp. xviii, 43. *Cambridge Greensand ; Cambridge.* [Vertebræ.]

Syngonosaurus macrocercus, H. G. Seeley, Q. J. G. S. xxxv (1879), p. 621, woodcuts 6—8. *Cambridge Greensand ; Cambridge.* [Vertebræ.]

Teleosaurus eucephalus, v. Pelagosaurus brongniarti.

Testudo cantabrigiensis, H. G. Seeley, Index to Aves, etc., Woodw. Mus. (1869), pp. xix, 32. *Cambridge Greensand ; Cambridge.* [Mandible.]

Thalassemys hugii, Rütimeyer. *Enaliochelys chelonia*, H. G. Seeley, Index to Aves, etc., Woodw. Mus. (1869), p. 108. *Kimeridge Clay ; Ely.* [Bones.]

Trachydermochelys phlyctœnus, H. G. Seeley, Index to Aves, etc.,
Woodw. Mus. (1869), pp. xix, 35. *Cambridge Greensand;
Cambridge.* [Fragments of carapace.]

Axis of Dinosaur' (*Iguanodon ?*), H. G. Seeley, Q. J. G. S. XXXI
(1875), p. 461, woodcuts 1, 2. *Wealden; Brook, Isle of Wight.*

AVES.

Argillornis longipennis, Owen. ? *Megalornis emuinus* (Bowerbank), H. G. Seeley, A. M. N. H. ser. 3, XVIII (1866), p. 110,
and Q. J. G. S. XXX (1874), p. 708, woodcut, f. 1. *London
Clay; East Chvrch.* [Shaft of limb-bone.]

Enaliornis barretti (Seeley), H. G. Seeley, Index to Aves, etc.,
Woodw. Mus. (1869), p. xvii, and Q. J. G. S. XXXII (1876),
pl. xxvi, f. 1—4, 7—22, 24—27, pl. xxvii, f. 4, 5, 12, 19, p. 496.
Palæocolyntus barretti, H. G. Seeley, Proc. Camb. Phil. Soc. I
(1864), p. 228 [misprint]. *Pelagornis barretti*, H. G. Seeley,
A. M. N. H. ser. 3, XVIII (1866), p. 110. *Palæocolymbus
barretti*, H. G. Seeley, Q. J. G. S. XXXII (1876), p. 497, footnote. 'Bird about the size of a woodcock,' R. Owen, Palæontology, second edition (1861), p. 327. *Cambridge Greensand;
Cambridge.* [Bones.]

E. sedgwicki (Seeley), H. G. Seeley, Index to Aves, etc., Woodw.
Mus. (1869), p. xvii, and Q. J. G. S. XXXII (1876), pl. xxvi,
f. 12, 13, pl. xxvii, f. 6—11, 13—18, p. 496. *Pelagornis
sedgwicki*, H. G. Seeley, Proc. Camb. Phil. Soc. I (1864), p. 228,
[misprinted *Pelargonis*]. *Cambridge Greensand; Cambridge.*
[Bones.]

Macrornis tanaupus, H. G. Seeley, A. M. N. H. ser. 3, XVIII
(1866), p. 110. *Lower Headon Beds; Hordwell.* [Proximal
end of right tibia.]

Megalornis emuinus, v. Argillornis longipennis.

Palæocolymbus barretti, v. Enaliornis barretti.

Palæocolyntus barretti, v. Enaliornis barretti.

Pelagornis barretti, v. Enaliornis barretti.

AVES. 179

Pelagornis sedgwicki, v. Enaliornis sedgwicki.

Ptenornis, H. G. Seeley, A. M. N. H. ser. 3, XVIII (1866), p. 109.
Hamstead Beds; Hamstead. [Portion of coracoid.]

MAMMALIA.

Bos taurus, Linnæus, var. **primigenius**, Bojanus. *B. primi-genius*, J. Carter, G. M. dec. 2, I (1874), p. 492. S. H. Miller and S. B. J. Skertchly, The Fenland (1878), p. 321, plate (left figure). J. E. Harting, Brit. Anim. Extinct (1880), p. 214. C. C. Babington, Comm. Camb. Antiq. Soc. II (1863), p. 285, plate. *Peat; Burwell.* [Skull, vertebræ, etc.]

Diplopus aymardi, W. Kowalevsky, Phil. Trans. CLXIII, 1873 (1874), pl. xxxviii, f. 8, 8′, p. 51. *Lower Headon Beds; Hordwell.* (This is the locality given by Kowalevsky, but the specimen is probably from Hamstead.) [Unciform.]

Elephas antiquus, Falconer, A. L. Adams, Brit. Foss. Elephants (1881), p. 178. H. Falconer, Pal. Memoirs, II (1868), p. 182. *Pleistocene; Whittlesea.* [Portion of mandible.]

E. meridionalis, Nesti, A. Leith Adams, Brit. Foss. Elephants (1881), pl. xxv, f. 1, 1 *a*, *b*, p. 202. *Forest Bed; Cromer.* [Mandible.] Presented by Miss Gurney.

E. primigenius, Blumenbach, A. L. Adams, Brit. Foss. Elephants (1879), pl. xiii. f. 1, 1 *a*, p. 111. *Pleistocene; Kirby, near Melton Mowbray.* [Molar.]

Specimens from the following localities are briefly described by A. L. Adams, *loc. cit.*:—Barton, Cambs, p. 95; Kirby, pp. 95, 100, 113; Lawford, near Rugby (No. 26) p. 96; Chesterton, p. 97 (and Falconer, Pal. Mem. II. (1868), p. 162); Gristhorpe Bay, p. 99; Cambridge, p. 99; Westwick Hall, p. 100; St Neots, pp. 100, 107; Lexden, pp. 100, 124 (Fisher Coll.); Barnwell, p. 106; Newton, I. W., p. 110; Crayford, p. 111; Walton-on-the-Naze, p. 112; Norfolk Coast, p. 118 (and Falconer, II. p. 170); 'Valley of the Danube,' p. 119; Barrington, p. 120; Wurtemberg, p. 121; Bigbone Lick, Kentucky, p. 122; locality lost, pp. 107, 115, 116 (and Falconer, II. p. 174), 121 (and Falconer, II, p. 169).

Hippopotamus amphibius, Linnæus. *H. major*, Owen, P. Lake, G. M. dec. 3, II (1885), p. 318. *Pleistocene; Barrington, Cambs.* [Mandible.]

180 MAMMALIA.

Hippopotamus major, v. Hippopotamus amphibius.

Hyopotamus, small species, W. Kowalevsky, Phil. Trans. CLXIII 1873 (1874), pl. xxxviii, f. 9, 9', p. 51. *Hamstead Beds; Hamstead, Isle of Wight.* [Unciform.]

Merycochœrus leidyi, G. T. Bettany, Q. J. G. S. XXXII (1876), pl. xviii, f. 1, 2, p. 270. *Miocene; John Day's River, Oregon.* [Skull.] Presented by Lord Walsingham.

M. temporalis, G. T. Bettany, Q. J. G. S. XXXII (1876), pl. xvii f. 1—4, p. 269. *Miocene; John Day's River, Oregon.* [Portions of skulls.] Presented by Lord Walsingham.

Squalodon sp., E. Ray Lankester, Q. J. G. S. XXI (1865), pl. xi, f. 4, p. 221. *Red Crag; Suffolk.* [Tooth.]

Palœobalœna sedgwicki, v. Palæocetus sedgwicki.

Palæocetus sedgwicki (Seeley), H. G. Seeley, G. M. II (1865), pl. iii, f. 1, 2, p. 54. W. H. Flower, Nature, XXIX (1883), p. 170. R. Lydekker, Cat. Foss. Mamm. B. M. part V (1887), p. 31. *Palœobalœna sedgwicki*, H. G. Seeley, Proc. Camb. Phil. Soc. I (1864), p. 228. *Boulder Clay; Ely.* (? derived from the Oxford or Kimeridge Clay). [Cervical vertebræ.]

Printed in the United States
By Bookmasters